SpringerBriefs in Applied Sciences and Technology

For further volumes:
http://www.springer.com/series/8884

Khin Wee Lai · Eko Supriyanto

Detection of Fetal Abnormalities Based on Three Dimensional Nuchal Translucency

Springer

Khin Wee Lai
Department of Clinical Science
and Engineering
Universiti Teknologi Malaysia
Skudai-Johor
Malaysia

Eko Supriyanto
Clinical Engineering Research Laboratory
Universiti Teknologi Malaysia
Skudai-Johor
Malaysia

ISSN 2191-530X   ISSN 2191-5318   (electronic)
ISBN 978-981-4021-95-1   ISBN 978-981-4021-96-8   (eBook)
DOI 10.1007/978-981-4021-96-8
Springer Singapore Heidelberg New York Dordrecht London

Library of Congress Control Number: 2012949376

Printed on acid-free paper

Springer is part of Springer Science+Business Media (www.springer.com)

Contents

1 Introduction 1
    1.1 Background of Research 1
    1.2 Problem Statement 6
    1.3 Objectives 6
    1.4 Scope of Project 7
    1.5 Contribution of Book 7
    References 8

**2 Medicine and Engineering Related Researches
    on the Utility of Two Dimensional Nuchal Translucency** 11
    2.1 Review of Trisomy 21 11
    2.2 History of Trisomy 21 Detection 12
    2.3 Maternal Age Factor 13
    2.4 Effect of Previous Pregnancies Factor 14
    2.5 Existing Detection Methods of Trisomy 21 15
        2.5.1 Ultrasound 15
        2.5.2 Maternal Serum Markers 17
        2.5.3 Genetic Testing 17
        2.5.4 Summary 18
    2.6 Two Dimensional Nuchal Translucency 21
        2.6.1 Medicine Review and Related Researches 21
        2.6.2 Engineering Review and Related Researches 24
    2.7 Three Dimensional Ultrasound Applications 37
        2.7.1 Summary 39
    References 41

**3 Designs and Implementation of Three Dimensional
    Nuchal Translucency** 47
    3.1 Introduction 47
    3.2 Data Acquisition 49
    3.3 DICOM and Clips Storage 52
    3.4 3D Nuchal Translucency Reconstruction and Visualization ... 55

3.4.1   Three Dimensional Scanning Acquisitions   60
3.4.2   3D Volumetric Rendering.   66
3.5   Three Dimensional Nuchal Translucency Segmentation   75
3.5.1   Pre-Processing 3D Diffusion   78
3.5.2   Three Dimensional Seeded Region-Based Segmentation . . .   84
3.5.3   Three Dimensional Nuchal Translucency
          In Vivo Measurement   90
References.   93

4   Clinical Tests and Measurements.   95
4.1   Introduction   95
4.2   Descriptive Analysis.   96
4.3   Intra Observer Variability   99
4.4   Level Agreement Between Measurements   101
4.4.1   Paired Sample t-Test.   104
4.5   Methods Repeatability   104
4.6   Summary   106
4.7   Discussion   106
4.7.1   Advantages and Disadvantages of Early Fetal
          Abnormalities Detection.   107
References.   108

5   Future Improvements   109
5.1   Conclusions   109
5.2   Limitations and Recommendations   110
5.2.1   The Need of Scanning Systems Improvement   110
5.2.2   The Need of Image and Position for Freehand
          Acquisition Improvement.   111
5.2.3   The Need of Algorithms for Fully
          Automated Segmentation   111
References.   111

Glossary   113

Symbols and Abbreviations

2D	Two dimensional
3D	Three dimensional
ACR	American College of Radiology
AFP	Alpha feto-protein
BMTI	Institute of Biomedicine Technique and Informatics
BPD	Biparietal diameter
CRL	Crump rump length
CTF	Color transfer function
CVS	Chorionic villus sampling
DIA	Inhibin A
DICOM	Digital Imaging and Communications in Medicine
DP	Dynamic programming
FMF	Fetal Medicine Foundation
GA	Gestation age
HUSM	Hospital Universiti Sains Malaysia
Hz	Frequency
ICOV	Instantaneous coefficient of variation
JEPeM	Ethical Committee (Human)
MA	Maternal age
MC	Marching cube
MPR	Multi-planar reformatting
MRI	Magnetic resonance imaging
NB	Nasal bone
NEMA	National Electrical Manufacturers Association
NT	Nuchal translucency
NTD	Neural tube defects
O&G	Obstetrics and gynecology
OTF	Opacity transfer function
PAPP-A	Pregnancy associated plasma protein A
PUBS	Percutaneous umbilical cord blood sampling
QF-PCR	Quantitative fluorescent polymerase chain reaction
ROI	Region of interest
SPL	Spatial pulse length

SRAD	Speckle reduction anisotropic diffusion
SRG	Seeded region growing
SRI	Speckle reduction imaging
uE3	Unconjugated oestriol
UK	United Kingdom
US	Ultrasound
UTM	Universiti Teknologi Malaysia
λ	Wavelength
v	Velocity
$G(i, j)$	Gradient magnitude
$\vartheta(i, j)$	Gradient phase
$C(x, y)$	Diffusion coefficient
B_n	Polylines for NT border
P_i	Neighboring pixels
$C(B_N)$	Cost function
W_i	Weight factor
r	Reference point
f	Frame rate
s	Acquisition time
N_i	Normal vector of vertex
V_i	Vertex of triangle
T_{3D}^x	Rotational matric transformation
T	Thresholding level
q_0	Initial speckle computed homogenous region
∇	Gradient operator
$E\left(\vec{pq}\right)$	Euclidean vector
$N(x, y, z)$	Neighborhood pixels
R_i	Seed regions
$H(.)$	Degree of region homogeneity
∂	Empirical threshold
$g(n)$	Image input
∇f	Strength of edge pixels
$\beta(x, y)$	Vector direction of edge

Chapter 1
Introduction

Abstract The research begins with the studies of fetal abnormalities related Trisomy 21 with maternal and fetal health data. An exploration on current modalities for Trisomy 21 detection methods namely Biochemistry triple tests, Genetic testing, and Ultrasound prenatal screening shall be completed prior to investigation of each of their manual working principle, advantages and drawbacks. With the help of clinician, ultrasound prenatal screening was chosen based on its clinical acceptance, cost, potential of usage and safe imaging. This existing imaging modality is used to formulate a new method for Trisomy 21 early detection due to its current critical limitations. Studied ultrasound markers are decided based on the result from previous researches, where nuchal translucency or NT is found to be most effective marker for Trisomy 21 assessment. Ultrasound fetal images and other related maternal materials will be collected from some sources, i.e. from online library, clinics and hospitals. They will be grouped and analysed, i.e. image history, image structure, noise and image extraction.This chapter includes an introduction, the background, the objective and scope of the book. The main aim is to show the motivation of this research and the existing limitation on ultrasound prenatal scanning protocol for Nuchal translucency. It will also describe its research methodology in brief. This chapter is summarized with the novelties and contribution of book and its feasibility.

1.1 Background of Research

Fetal abnormalities can be detected through the study of particular ultrasound markers for example like nuchal translucency (NT), nasal bone, length of femur and ductus venous (Nicolaides et al. 1992; Snijders et al. 1998; Zosmer et al. 1999). Recent studies show that these ultrasound markers can be used for identification of incidence chromosomal abnormalities, such as Trisomy 21 (Hyett et al. 1995a, b; Souka et al. 2001). Trisomy 21 (British Down's syndrome) is a genetic condition where by an extra copy of 21st chromosome results in abnormalities

K. W. Lai and E. Supriyanto, *Detection of Fetal Abnormalities Based on Three Dimensional Nuchal Translucency*, SpringerBriefs in Applied Sciences and Technology, DOI: 10.1007/978-981-4021-96-8_1, © The Author(s) 2013

in fetus development. It will, consequently, deviate the development of normal physical body structure, functions, and it often leads to mental retardation. Other examples of Trisomy include Trisomy 18 and Trisomy 13 (Tul et al. 1999; Celentano et al.2003); Trisomy 18 or Trisomy 13 is a chromosomal abnormality: three copies of the 18th chromosome (or of the 13th chromosome) present in each cell of the body instead of the usual pair. Trisomy 21 or Down's syndrome, perhaps, is one of the most frequent congenital causes of severe mental retardation with an incidence at birthrate 1.3 per 1000 (Hulten et al. 2010).

Currently, the presence of Trisomy 21 can be detected manually by using three different methods; biochemistry blood tests, prenatal ultrasound screening and genetic confirmatory testing. The verified blood biochemistry markers for screening the Trisomy 21 are the so called triple test of mother's blood, which includes Pregnancy Associated Plasma Protein A (PAPP-A), free beta human chorionic gonadotropin (β-hCG) and maternal alpha-fetoprotein (AFP). Based on previous researches, it has been recognized that the chromosomally abnormal pregnancy is associated with the abnormal level of biochemical markers. AFP is produced by fetus while PAPP-A and free ß-hCG are produced by placental trophoblast during pregnancy (Wee et al. 2010a, b). In the first trimester, the PAPP-A level is, on average, low in Down's syndrome pregnancies (about half that of unaffected pregnancies). In the second trimester AFP levels is, on average, low (about three-quarters that of unaffected pregnancies) and free ß-hCG levels is, on average, high (about double that of unaffected pregnancies). Nevertheless, the drawbacks of these biochemistry tests are low cost effectiveness, invasiveness and time consumption. The common practice nowadays is, combining biochemistry markers with ultrasound markers for higher reliability assessment.

Genetic testing is categorized into invasive method such as amniocentesis, chorionic villus sampling (CVS), or percutaneous umbilical cord blood sampling (PUBS). Only well trained or experienced O&G specialists are permitted to execute the collection of biopsy through these invasive methods above. Due to its critical drawbacks in term of time usage, cost and potential miscarriage risk, it has always been considered as non-preferred technique during premature fetal screening and only been regarded for confirmatory testing at the last stage of clinical abnormalities screening (Wee and Supriyanto 2010).

Ultrasound screening in first trimester of pregnancy provides the most effective way of chromosomal abnormalities screening. Previous researches show that assessment of particular ultrasound markers offers promising non-invasive method for fetal abnormalities detection: nuchal translucency, nasal bone, long bone biometry, maxillary length, cardiac echogenic focus and ductus venous (Nicolaides et al. 1994; Cicero et al. 2003). Nevertheless, the drawback of current ultrasound manual measurement technique is restricted with inter and intra-observer variability and inconsistency of results (Pandya et al. 1995; Abuhamad 2005). Besides, the utility of two-dimensional image plane invariably depends on the physical structure of the region of interest. Of most of the existing medical imaging system, it is not trivial to generate the best spatial orientation two-dimensional image directly (Lai Khin, Arooj et al. 2010; Lai Khin and Supriyanto 2010; Lai Khin, Too Yuen et al. 2010; Supriyanto, Lai Khin et al. 2010); the structures of positioning and scanning orientation are generally subject to the structure and other physical limitations.

In many occasions, it is crucial to identify and display the best two dimensional (2D) image planes. For instance, the best sagittal plane during the prenatal ultrasound screening is favorable for physician to access certain ultrasound markers: nuchal translucency, nasal bone, or ductus venous for Trisomy 21 assessment. This is of prime significance since current manual scanning method is restricted to acquire the correct scanning plane of 2D ultrasound fetal images; if the ultrasound marker is not examined in an appropriate plane, the measurement would be shorter or longer than normal or in worst case — it does not exist. Also, if the tested images are not in the true sagittal view or coincide in the suitable plane, ultrasound markers might not appear in appropriate position. This difficulty remains unsolved in a few cases; further explanation can be seen at Sect. 3.2.

NT measurements were performed according to the standards established by Nicolaides et al. (1992). The competency of ultrasound operator worldwide for NT measurement are verified by the examination conducted by Fetal Medicine Foundation (FMF), led by Nicolaides et al. based in United Kingdom (UK). Healthcare professionals who passed the examination will be given certified licenses, and on-going audits to recertify their licenses are performed every successive year. Up to year 2010, the FMF database shows the number of registered certified healthcare personnel from Malaysia is only ten, while Germany's registered medical personnel are 502 persons. Figure 1.1 shows the number of doctors who passed NT measurement competency in some countries for comparison. Apparently Asian countries, i.e. Malaysia do not show the trend of high number of certified holders

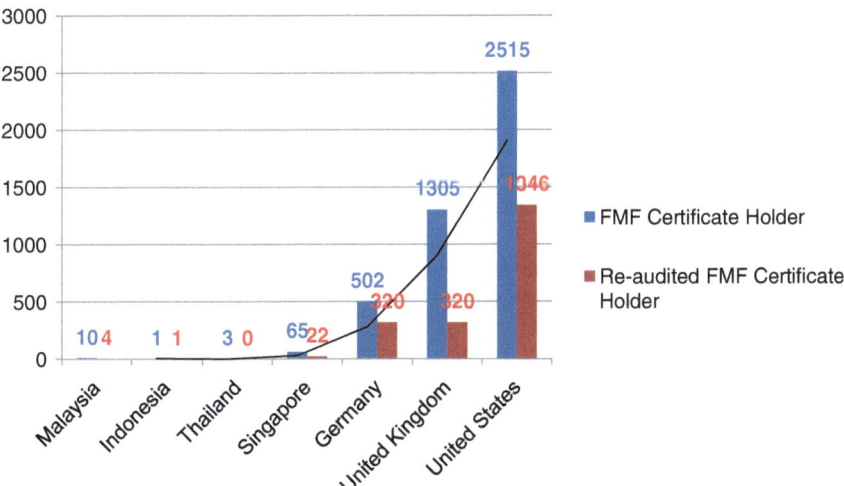

Fig. 1.1 Holder of the FMF certificate in the measurement of nuchal translucency in various countries

Table 1.1   Number of doctors in Malaysia and its ratio in population

Years	Number of doctors	Population: health personnel ratio
1990	7,012	2569:1
2000	15,619	1413:1
2007	23,738	1145:1

Sources Malaysia (1996), (2008)

Table 1.2   Distribution of doctors by State, Malaysia for year 2005

State	Number of doctors						
	Public sector			Private doctors	% of private doctors	Total number. of doctors	Ratio of population to doctors
	MOH	Non-MOH	Total				
Johor	832	12	844	885	51.2	1,729	1794:1
Kedah	529	6	535	452	45.8	987	1872:1
Kelantan	378	377	755	188	19.9	943	1596:1
Melaka	315	26	341	337	49.7	678	1051:1
N. Sembilan	464	6	470	324	40.8	794	1191:1
Pahang	483	1	484	315	39.4	799	1786:1
Perak	661	24	685	811	54.2	1,496	1509:1
Perlis	98	1	99	37	27.2	136	1655:1
Penang	665	9	674	851	55.8	1,525	963:1
Sabah	754	3	757	352	31.7	1,109	2719:1
Sarawak	722	25	747	366	32.9	1,113	2078:1
Selangor	962	93	1,055	2,078	66.3	3,133	1512:1
Terengganu	328	0	328	146	30.8	474	2145:1
Kuala Lumpur	1,177	948	2,125	1,801	45.9	3,926	396:1
Total	8,368	1,531	9,899	8,943	47.5	18,842	1387:1

Note MOH (Ministry of Health, Malaysia), *Source* Ninth Malaysia Plan, Table 20-5

for NT measurement. Consequently, measurement confidence and reliability are not assured. This condition could be worst with high ratio of doctors to patient population although this condition have improved over the last decade. Table 1.1 shows the population statistics of health personnel ratio for year 1990, 2000, and 2007 in Malaysia.

The imbalanced distributions of doctors among rural and urban areas in Malaysia are shown in Table 1.2. This can be seen for Kuala Lumpur with the best ratio 396:1; while Sabah has highest ratio (or worst ratio) with 2719:1. Another factor that could also contribute for this problem is the availability of ultrasound machine in hospitals especially in rural places, which is influenced by the number of patient in a hospital, as shown in Table 1.3. In particular, public hospitals have maximum of 17 ultrasound machines only if the patient capacity is more than 500. This means that only capital cities with large general hospital can have high number of ultrasound machine.

Table 1.3 Number of US machines in Malaysian hospitals

Capacity of hospital	Number of hospitals		Average number. of in-patients admission		Maximum number. of US machines	
	Public	Private	Public	Private	Public	Private
>5–50	4	13	3,430	1,347	2	4
51–100	8	12	6,783	6,229	4	5
101–500	19	15	20,899	16,309	12	15
>500	13	0	49,432	–	17	–

Note Capacity above is based on bed-count, *Sources* Healthcare Survey 2009

Based on the statistics above; extreme low number of FMF certified medical doctors, high ratio of doctors to population, and low number of available ultrasound system, provide good conventional 2D B-mode NT marker assessment based on gold standard developed by FMF while it is inherent with existing limitations which seems to be highly unrealizable. 3D NT marker assessment will be a breakthrough in this research for solving this problem, however, not to mention that the high purchasing costs of 3D US systems and its available low number of machines in Malaysia hospitals, it is also intolerable to perform 3D NT marker using commercial 3D US system due to technical design problem (refer to section Data acquisition), and it is not covered in routine prenatal scanning protocol.

Therefore, in our case, three dimensional ultrasound fetal image reconstruction and visualization has great significance in applications: the diagnosis of 3D fetus, scene planning and 3D simulation, 3D Nuchal translucency assessment, 3D Nuchal translucency segmentation and its interactive measurement. 3D ultrasound imaging has higher value for clinical application. The motivation of the book is resolving the entire current limitations of ultrasound marker measurement protocol, by replacing and upgrading the existing 2D ultrasound system to higher level—3D assessment without extra hardware modification and costs, while reducing human intervention, avoiding incompetent operator, to ensure the conducted measurement is less reliant on human skills through the use of automated computing system.

There are five major expected contributions, firstly, the new method for 3D NT measurement is less reliant on human intervention and operator dependent; second, reconstructed 3D image has higher clinical values for diagnostics as compared to conventional B-mode image; third, limitation of true mid-sagittal plane selection can be resolved through 3D NT assessment; forth; 3D NT measurement as higher repeatability performance; lastly, visualization and measurement of proposed 3D method is not restricted to fetus position. In summary, early detection of fetal abnormalities during the first trimester of pregnancy is important, and early detection enables improvement of mothers' heath quality. It may safe mother's life due to fetal abnormalities that can be dangerous. Besides, the parents may have early information regarding their baby, so they can be better prepared to receive the child.

1.2 Problem Statement

The critical limitations of the existing NT ultrasound manual scanning method are operator dependent, observer variant and inconsistency of measurement result due to improper training, human error, operators' expertise and experience. Current clinical practice of manual NT examination method by locating the sonogram calipers on 2D Ultrasound image requires highly trained and experienced operators by adhering to a standard tedious protocol, and is therefore prone to errors, intra-observer and inter-observer repeatability can be questioned. Moreover, it is expected the manual examination will cause the problem of drift in measurements over time in longitudinal studies.

True sagittal plane selection is another existing unsolved clinical practice limitation. Different selected plane position will result in different size of interest marker in 2D image formation. Medical personnel using hand-eye coordination process to build up a detailed mental picture of the 3D anatomy is subjected to their experience and skill. Mean of two or three trials best 2D ultrasound marker measurement is considered as final parameter to assess the Trisomy risk. Admittedly, reproducibility and efficiency cannot be guaranteed under these circumstances.

Competency of operator for NT measurement could be one of the doubtful difficulties, if the particular healthcare personnel do not take the FMF examination, the measurement may not be reliable. Underestimation or overestimation of NT fold thickness is therefore inherent in their measurements. As the number of FMF registered doctor in Malaysia is very low, i.e. ten doctors, less people can benefit from this technology. Hence, a new method for 3D NT assessment needs to be developed with the aim of improving the inefficiency and reducing the tedious manual examination procedure.

1.3 Objectives

Based on the needs described in the last section, an open-source 3D Nuchal Translucency ultrasound marker reconstruction and assessment using conventional hardware equipment shall be developed. The book aims to develop new algorithms so as to achieve the four objectives outlined in this section;

- 3D ultrasound reconstruction and visualization based on conventional 2D ultrasound imaging system
- 3D semi-automated segmentation based on physiological structure and characteristics of Nuchal translucency
- Interactive visualization of internal ultrasound marker—3D Nuchal translucency
- High flexibility volume thickness measurement of 3D Nuchal translucency formation.

The conventional 2D ultrasound imaging system mentioned in the first objective above is aim to provide an open-source platform for 3D ultrasound reconstruction without extra costs. The intention of second objective is to provide an intelligent automated method for 3D NT segmentation that are less reliant on human intervention. The third and fourth objectives are proposed in order to design and develop a highly flexible and interactive visualization mode for internal ultrasound marker of 3D NT.

1.4 Scope of Project

There are few scopes to ensure the study is conducted within the boundary set, and heading in the right direction to achieve the intended objectives. This research will focus on the acquisition of 2D ultrasound fetal images during first trimester of gestation period. The population of the collected data are merely Malaysian which includes Malay and Chinese mainly. Maternal ages range between 25 and 40 years old. Only singleton pregnancies are included in this research with specific ultrasound marker investigation, namely Nuchal Translucency (NT). Conventional 2D B-mode ultrasound machine with acceptable resolution are applied with trans-abdominal type ultrasound probe at frequency 3.5 MHz. Ethical approval for data collection at the hospital was granted by the ethical committee Hospital Universiti Sains Malaysia (HUSM). Total number of hospital patients is 23.

Methods assumption for multiple 2D successive ultrasound frame acquisition is in linear position with respect to constant speed of transducer movement. This has to be coordinated with experienced sonographer in advance. The scan areas will be cross-sectioned NT fold thickness region in sagittal view mode. Recording range of moving slices shall be small with prerequisites; maximum thickness of NT must be enclosed.

1.5 Contribution of Book

To the best knowledge of the author, the proposed open-source 3D NT reconstruction and assessment algorithm introduces the following list of novelties in the field: It is the first open-source freehand 3D Nuchal Translucency ultrasound reconstruction and interactive visualization. It is also the first 3D Nuchal Translucency assessment system that employs only conventional 2D B-mode ultrasound machine and non-customized hardware. Besides, this work also contributed a new method for 3D NT measurement function equipped with intelligent semi-automated 3D NT segmentation technique.

This research is aimed at converting the proof-of-principle system from conventional ultrasound 2D B-mode NT measurement into a clinically viable 3D NT assessment using open-source approach. Results findings are not restricted to the physiological obstacle or ultrasound transducer position. Free orientation virtual slider in 360° was designed in various inspection view mode, i.e. 3D full volume rendering, multiplanar reformatting (MPR) and hybrid slider view. It is expected reconstructed 3D data are more clinical enlightening and have higher clinical values as compared to conventional 2D images. Repeatability of measurement is promising while patients do not have to perform another new scanning procedure again.

Existing 2D NT examination protocol are blinded to the image position on 3D anatomy. It is mapped on the virtual 3D structure based on medical personnel's imagination and their understanding on the interest organ structure. Conventional ultrasound images are 2D, yet the anatomy of nuchal translucency is 3D, hence the diagnostician must integrate multiple images in his mind. This practice is inefficient, and may lead to variability and incorrect diagnoses. Besides, the 2D ultrasound image represents a thin plane at some arbitrary angle in the body. It is difficult to localize the image plane and reproduce it at a later time for follow-up studies. Within the proposed technique in this book, it shall be replaced by 3D volume rendering and false sagittal plane with maximum NT thickness can be avoided easily. The studied marker NT will be shown as explicit 3D 'valley' rather than two weak echogenic lines on 2D B-mode ultrasound images. Hence, overestimation or underestimation can be avoided and therefore, result consistency and accuracy can be ensured.

Progress has been made by developing new measurement methods and algorithms for 3D NT formation with high interactive visualization capability, while no changes have been made to the conventional scanning hardware at extra costs. This is essentially important for developing countries like Malaysia, where conventional ultrasound machine were employed in most of the hospital rather than 3D high end ultrasound machine with expensive purchasing and maintenance costs. Consequently, higher number of small-scaled hospitals can afford this system and larger public community population can be benefited.

This work is strictly technical, and its emphasis was influenced by the opinions of clinical collaborators. Resources and ethical approval for clinical scanning of hospital patients was granted during the period of technical research. Attention was paid to making the 3D NT assessment system feasible, easy to use and replaced tedious manual measurement protocol, improved consistency, reduced human intervention and operator dependency, avoided competency factor and human errors, while producing reliably meaningful images and measurement, so as to support future studies in a clinical setting.

References

Abuhamad, A. (2005). Technical aspects of nuchal translucency measurement. *Seminars in Perinatology, 29*(6), 376–379.

Celentano, C., Di Donato, N. G., Prefumo, F., & Rotmensch, S. (2003). Early resolution of increased nuchal translucency in a fetus with trisomy 18. *American Journal of Obstetrics and Gynecology, 189*(3), 880–881.

Cicero, S., Longo, D., Rembouskos, G., Sacchini, C., & Nicolaides, K. H. (2003). Absent nasal bone at 11–14 weeks of gestation and chromosomal defects. *Ultrasound in Obstetrics and Gynecology, 22*(1), 31–35.

Hyett, J. A., Moscoso, G., & Nicolaides, K. H. (1995a). Cardiac defects in 1st-trimester fetuses with trisomy 18. *Fetal Diagnosis and Therapy, 10*(6), 381–386.

Hyett, J. A., Moscoso, G., & Nicolaides, K. H. (1995b). First-trimester Nuchal Translucency and cardiac septal-defects in fetuses with Trisomy-21. *American Journal of Obstetrics and Gynecology, 172*(5), 1411–1413.

Hulten, M., Patel, S., Westgren, M., Papadogiannakis, N., Jonsson, A., Jonasson, J., et al. (2010). On the paternal origin of trisomy 21 down syndrome. *Molecular Cytogenetics, 3*(1), 4.

Malaysia (1996). *Seventh Malaysia Plan 1996–2000.* Kuala Lumpur: Percetakan Nasional Malaysia.

Malaysia (2008). *Kajian Separuh Penggal Rancangan Malaysia Kesembilan 2006–2010. Putrajaya*: Unit Perancang Ekonomi, Jabatan Perdana Menteri.

Nicolaides, K. H., Azar, G., Byrne, D., Mansur, C., & Marks, K. (1992). Fetal nuchal translucency: ultrasound screening for chromosomal defects in first trimester of pregnancy. *British Medical Journal, 304*(6831), 867–869.

Nicolaides, K. H., Brizot, M. L., & Snijders, R. J. M. (1994). Fetal nuchal translucency—ultrasound screening for fetal trisomy in the first trimester of pregnancy. *British Journal of Obstetrics and Gynaecology, 101*(9), 782–786.

Pandya, P. P., Altman, D. G., Brizot, M. L., Pettersen, H., & Nicolaides, K. H. (1995). Repeatability of measurement of fetal nuchal translucency thickness. *Ultrasound in Obstetrics and Gynecology, 5*(5), 334–337.

Snijders, R. J. M., Noble, P., Sebire, N., Souka, A., Nicolaides, K. H., & Grp, F. M. F. F. T. S. (1998). Uk multicentre project on assessment of risk of trisomy 21 by maternal age and fetal nuchal-translucency thickness at 10–14 weeks of gestation. *Lancet, 352*(9125), 343–346.

Souka, A. P., Krampl, E., Bakalis, S., Heath, V., & Nicolaides, K. H. (2001). Outcome of pregnancy in chromosomally normal fetuses with increased nuchal translucency in the first trimester. *Ultrasound in Obstetrics and Gynecology, 18*(1), 9–17.

Supriyanto, E, Wee L. K., Min, T. Y. (2010). Ultrasonic marker pattern recognition and measurement using artificial neural network. *Proceedings of the 9th WSEAS international conference on Signal processing,* 35–40.

Tul, N., Spencer, K., Noble, P., Chan, C., & Nicolaides, K. (1999). Screening for trisomy 18 by fetal nuchal translucency and maternal serum free beta-hcg and Papp-A at 10–14 weeks of gestation. *Prenatal Diagnosis, 19*(11), 1035–1042.

Lai Khin, W., & Supriyanto, E. (2010). Automatic detection of fetal nasal bone in 2 dimensional ultrasound image using map matching. *Proceedings of the 12th WSEAS International Conference on Automatic Control, Modelling & Simulation (ACMOS 2010), Catania, Italy* (pp. 305–309).

Wee, L. K., Lim, M., & Supriyanto, E. (2010a). Automated risk calculation for trisomy 21 based on maternal serum markers using trivariate lognormal distribution. *Proceedings of the 12th WSEAS International Conference on Automatic Control, Modelling & Simulation (ACMOS 2010), Catania, Italy* (pp. 327–332).

Wee, L. K., Miin, L., & Supriyanto, F. (2010b). Automated trisomy 21 assessment based on maternal serum markers using trivariate lognormal distribution. *WSEAS Transactions on Systems, 9*(8), 844–853.

Lai Khin, W., Too Yuen, M., Arooj, A., & Supriyanto, E. (2010). Nuchal translucency marker detection based on artificial neural network and measurement via bidirectional iteration forward propagation. *WSEAS Transactions on Information Science and Applications, 7*(8), 1025–1036.

Lai Khin, W., Arooj, A., & Supriyanto, E. (2010). Computerized automatic nasal bone detection based on ultrasound fetal images using cross correlation techniques. *WSEAS Transactions on Information Science and Applications, 7*(8), 1068–1077.

Zosmer, N., Souter, V. L., Chan, C. S. Y., Huggon, I. C., & Nicolaides, K. H. (1999). Early diagnosis of Major cardiac defects in chromosomally normal fetuses with increased nuchal translucency. *BJOG: An International Journal of Obstetrics & Gynaecology, 106*(8), 829–833.

Chapter 2
Medicine and Engineering Related Researches on the Utility of Two Dimensional Nuchal Translucency

Abstract A thorough literature review on current research for trisomy 21 detection using ultrasound will be carried out for existing modality's drawback investigation. Due to its critical restriction, computing on ultrasound markers in term of its recognition, segmentation and measurement are essentially required. 3D reconstruction of nuchal translucency becomes a breakthrough to select the appropriate scanning plane of ultrasound markers. It shall resolve the problematic issue on scanning plane selections which depends on operator assumption and experience. Data sources from hospital patient scanning should obtain the approval from Medicine ethical committee. Consent and simple agreement document shall be prepared. The main ultrasound images sources will be taken from Health Center Universiti Teknologi Malaysia, Hospital Universiti Sains Malaysia, and Hospital Universiti Kebangsaan Malaysia. External collaboration parties are Hospital Sultanah Aminah and Technische Universitat Ilmenau, Germany. Other than fetal data, maternal health data will also be recorded. This data is important to obtain the knowledge of correlation between fetal and mother data. It is recommended that pregnant women should be older than 30 years old. This chapter will describe the book in terms of its background, history and the related works in greater detail. The focus will be on the Trisomy 21 background, history, existing detection techniques, and ultrasound application using 2D and 3D image formation on fetal abnormalities detection. Previous related research works are discussed and each of their limitation is remarked.

2.1 Review of Trisomy 21

Trisomy 21 or Down syndrome is the most common disease of chromosomal abnormalities, where the patients' cells have extra copy of 21st chromosomes as compared to normal paired chromosomes, leading to abnormal structure and function of many organs, including mental retardation, congenital heart disease, and intestinal plugs. Other examples of Trisomy include Trisomy 18 and Trisomy 13. Trisomy 18 or Trisomy 13 simply means there are three copies of the 18th chromosome (or of the 13th chromosome) present in each cell of the body, rather than the usual pair. It was firstly reported by Down in 1866 and it is named after him as Down syndrome (Down

K. W. Lai and E. Supriyanto, *Detection of Fetal Abnormalities Based on Three Dimensional Nuchal Translucency*, SpringerBriefs in Applied Sciences and Technology, DOI: 10.1007/978-981-4021-96-8_2, © The Author(s) 2013

1995). The main characteristics of this syndrome are severe mental retardation, with a unique facial and body deformities (Wee et al. 2010a, b).

The birthrate of Down's syndrome is approximately one in every 800–1000 live births. Affected babies are likely to suffer from severe mental and physical disabilities, affecting in particular the heart, gastrointestinal tract, eyes and ears. Down's syndrome generally lives to adulthood, but they need to receive long-term caregivers. In actual life, patient with trisomy 21 requires lifelong care and supports from their families, which will definitely cause heavy burden in both mental and economic wise.

2.2 History of Trisomy 21 Detection

In 1930s, Waardenburg and Bleyer were the first persons to speculate that the cause of Trisomy 21 might be due to chromosomal abnormalities. With the discovery of karyotype techniques in the 1950s, it became possible to identify abnormalities of chromosomal number or shape. In 1959, Jerome and Patricia were the first to determine the cause of Trisomy was due to the triplication of 21st chromosome. The chromosomes are thread-like structures composed of DNA and other proteins. They are present in every body cell and carry the genetic information needed for cell development. Genes, which are units of information, are encoded in the DNA. Normally, human cells have 46 chromosomes which can be arranged in 23 pairs. Of these 23, 22 are alike in males and females; these are called the autosomes. The 23rd pair is the sex chromosomes ('X' and 'Y'). Each member of a pair of chromosomes carries the same information, in that the same genes are in the same spot on the chromosome. However, variations of that gene may be present. For example, the genetic information for eye color is a "gene;" the variations for blue, green, etc.

Divisions of human cells are separated into two different ways. The first is ordinary cell division, which is also known as mitosis, by which the body grows. In this method, one cell becomes two cells which have the exact same number and type of chromosomes as the parent cell. The second method of cell division occurs in the ovaries and testicles, which is known as meiosis consisting of one cell splitting into two, with the resulting cells having half the number of chromosomes of the parent cell. So, normal eggs and sperm cells only have 23 chromosomes instead of 46. Figure 2.1 below shows the normal 23 pairs of chromosomal arrangement.

In Down syndrome, 95 % of all cases are caused by this event, where one cell has two 21st chromosomes instead of one, so the resulting fertilized egg has three 21st chromosomes. Hence the scientific name, Trisomy 21. Figure 2.2 below illustrate the structure of abnormal chromosomal arrangement.

The cause of trisomy 21 with an extra copy of chromosome is still unknown, but early researches have proved that it is highly associated with the maternal ages. Unfortunately, there are no effective prevention and treatment measures for this disease.

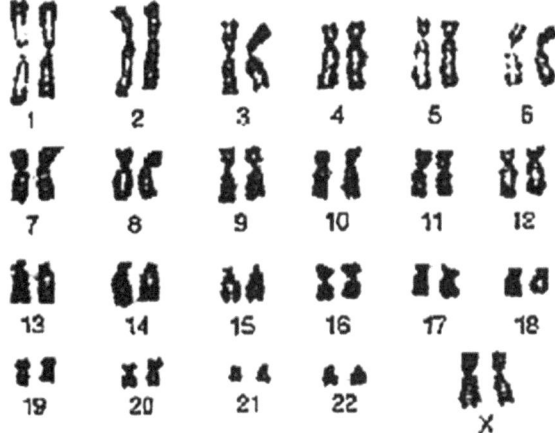

Fig. 2.1 Normal pairs of human chromosomal arrangement (Leshin 1997)

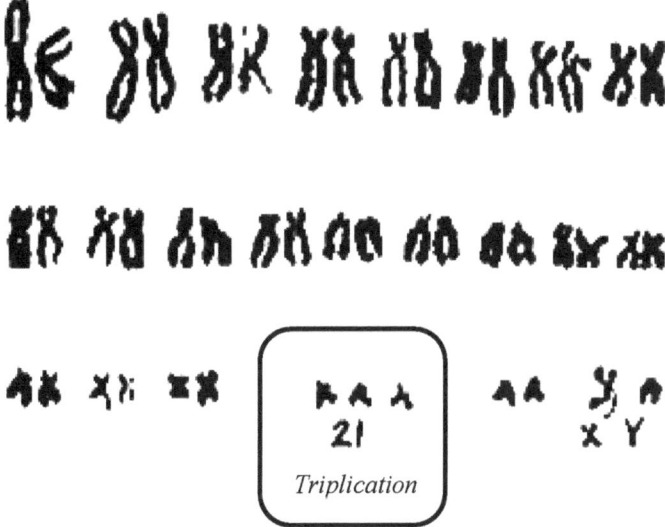

Fig. 2.2 Structure of abnormal chromosomal arrangement. *Note* XY means this is a karyotype of male with trisomy 21 (Leshin 1997)

2.3 Maternal Age Factor

Maternal age is the best known risk factor for trisomy 21 and other chromosomal abnormalities since 1980. The reasons why the age of the mother increases the risk for chromosomal abnormalities are still unknown currently. However, one of idea that is predicted by scientists is that older eggs are more prone to nondisjunction

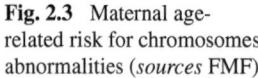

Fig. 2.3 Maternal age-related risk for chromosomes abnormalities (*sources* FMF)

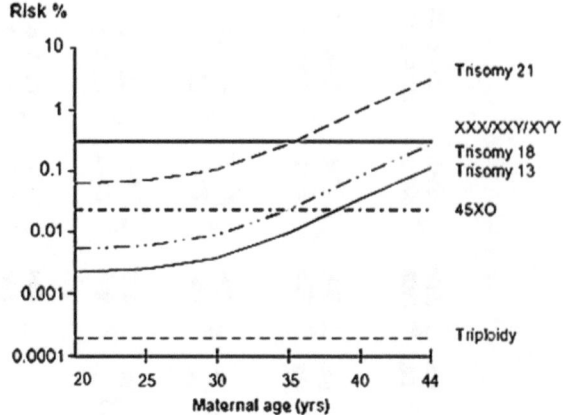

which in turn leads to the occurrence of trisomy 21, 18 and 13. For example, female eggs ovulated at age 40 have been in meiosis I for more than 40 years. During this time, events in the cell or environmental agents might damage the egg, making nondisjunction more likely.

Based on Huether (1998), maternal age highly influences the risk of conceiving Trisomy 21 baby. The statistics show that at maternal age in between 20 and 24, the probability is one in 1562; it increases dramatically to one in 214 at maternal age from 35 up to 39, and the probability for maternal with age 45 above is one in 19. The data shows the fact that elder maternal age has higher probability of conceiving Trisomy 21 baby.

The risk of Trisomy 21 and some other chromosomal abnormalities in an unborn child is known to increase with the age of the mother and it is this knowledge which forms the basis for selection of pregnant women for further investigation. Figure 2.3 shows the example of maternal age-related risk for chromosomes abnormalities. It can be observed that the increase of maternal age will have positive exponential risk increment of trisomy 21.

2.4 Effect of Previous Pregnancies Factor

Normally, couples who have one child with trisomies have a slightly increased risk of having a second child with trisomies. The recurrent risk of trisomies increase in current pregnancy because some couple with a previously affected pregnancy have parental mosaicism or a genetic defect that interferes with the normal process of disjunction (Snijders et al. 1999). In woman who had a previous pregnancy with trisomies, the risk of recurrent in the subsequent pregnancy is 0.75 % higher than maternal and gestational age-related risk for trisomies at the time of testing.

2.5 Existing Detection Methods of Trisomy 21

Basically, there are three different methods for Trisomy 21 detection during first and second trimester of pregnancy, including specific B-mode ultrasound marker assessment, maternal serum marker assessment and genetic examination of amniotic fluid or fetus blood. Brief description of each existing detection method in clinical practice are discussed as per below and comparisons of their pro and cons were conducted.

2.5.1 Ultrasound

There is extensive evidence (Nicolaides et al. 1992) that effective prenatal screening for major chromosomal abnormalities can be provided in the first trimester of pregnancy. Ultrasound screening in first trimester of pregnancy provides an effective way of screening chromosomal abnormalities (aneuploidy). Recent studies shows that assessment of particular ultrasound markers offer promising noninvasive method for fetal abnormalities detection, such as nuchal translucency, nasal bone, long bone biometry, maxillary length, cardiac echogenic focus and ductus venous (Nicolaides et al. 1999). American college of obstetricians and gynecologists has updated their guidelines and has recommended that all the pregnant women should be counseled about availability of screening tests for fetal aneuploidies (Acog 2007). By determining the risk in first trimester earlier reassurance for those with normal babies and safer termination for those with aneuploidy fetuses, is possible.

Medical literature has proven a fetus with congenital disease such as Trisomy 21, heart disease and bone disease will have thicker transparent layer of subcutaneous fetal neck or called Nuchal Translucency. The term Nuchal Translucency was termed by Nicolaides, pioneer in prenatal Trisomy 21 early assessment at Fetal Medicine Foundation (FMF), UK (Nicolaides et al. 1992). The formation of this transparent layer of skin was due to blockage of blood or lymphatic circulation, resulting in accumulation of liquid behind fetus's neck (Souka et al. 2001; Hyett et al. 1995; Kagan et al. 2009; Snijders et al. 1998). Single marker evaluation of NT can help doctors to evaluate the chances of fetal with Down syndrome up to 70 % (Abuhamad 2005; Zosmer et al. 1999). Some previous publication (Kagan et al. 2008; Cicero et al. 2003) used to assess the risk of Trisomy 21 during early pregnancy using NT measurement combining with pregnancy-induced plasma protein A (PAPP A) and free-Beta human chorionic gonadotropin (free ß-hCG), is able to drive the rate up to 90 % (one-stop reference trimester Down syndrome screening). Figure 2.4 shows the clinical anatomy of nuchal translucency formation.

With respect to the NT structure shown above, it can be examined using B-mode ultrasonic imaging by adhered to FMF guideline, as shown in Fig. 2.5 in

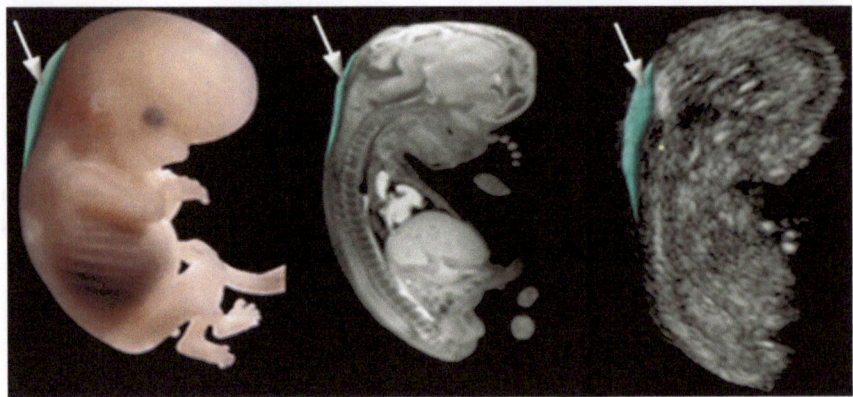

Fig. 2.4 Formation of nuchal translucency thicknesses due to fluid accumulation behind fetus' neck

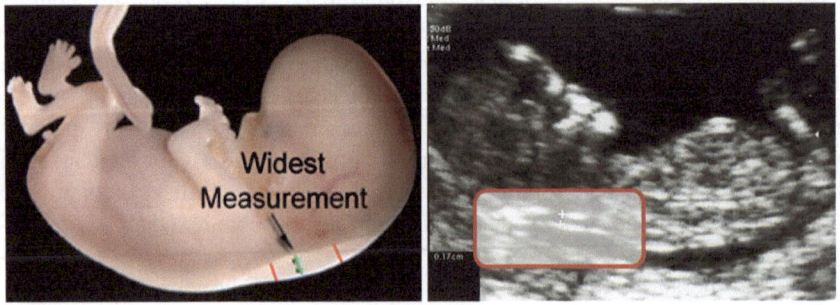

Fig. 2.5 Nuchal translucency thickness measurements for Trisomy 21 assessment during first trimester pregnancy

sagittal mode. NT thickness should be measured during the specific gestation ages, from 11 until 13 weeks plus 6 days, within Crump Rump Length (CRL) measured in between 45 and 84 mm. Only the maximum thickness of fold thickness regards as final measurement for Trisomy 21 assessment. Note that higher thickness of NT layer indicates the higher probability of Trisomy 21.

Maximum NT thickness measurement using B-mode ultrasonic imaging should strictly adhere to a tedious protocol developed by FMF. Although it provides a noninvasive method for Trisomy 21 assessment, there are some critical weaknesses of existing manual 2D marker measurement method, as discussed in Sect. 1.2. Details of medicine and technical engineering literature for ultrasound marker assessment are discussed in Sect. 2.7.1.

2.5.2 Maternal Serum Markers

Maternal serum markers are defined as a hormone or protein found in maternal blood that can serve as a sign of abnormality. The most common of these markers being alpha fetoprotein (AFP), pregnancy associated plasma proteins A (PAPP-A), unconjugated oestriol (uE3), free ß-human chorionic gonadotrophin (free ß-hCG) and inhibin A (DIA). It has been recognized that the chromosomally abnormal pregnancy is associated with the abnormal level of maternal serum markers. Both AFP and UE3 are produced by fetus while DIA, PAPP-A and free ß-hCG are produced by placental trophoblast during pregnancy (Wald et al. 1996).

In the first trimester, the PAPP-A level is, on average, low in Down's syndrome pregnancies (about half that of unaffected pregnancies) (Wald et al. 1996). In the second trimester AFP and uE3 levels are, on average, low (about three-quarters that of unaffected pregnancies) and inhibin- A and free ß-hCG levels are, on average, high (about double that of unaffected pregnancies). Kagan et al. (2008) had demonstrated that the maternal serum markers screening for calculation of accurate patient-specific risks for trisomy 21 is essential to take into account gestation age, maternal weight, ethnicity, smoking status and method of conception.

Before these biochemistry screening methods were introduced in the late 1980s, maternal ages are the single evaluation factor with the aim to select the 'high-risk' group of Trisomy 21. At 16 weeks of gestation, AFP, uE3 and hCG in Trisomy 21 pregnancies are sufficiently different from normal pregnancies to distinguish the risk group. This method improved the effectiveness than maternal age alone, identified about 50 until 70 % of the fetuses with Trisomy 21. After the emergence of ultrasonic marker NT in 1992 by Nicholaides et al. screening by a combination of maternal age and fetal NT thickness at 11–13 + 6 weeks of gestation was introduced. This method has now been shown to identify about 75 % of affected fetuses for a screen-positive rate of about 5 %. Subsequently, maternal age was combined with fetal NT and maternal serum biochemistry (free ß-hCG and PAPP-A) in the first-trimester to identify about 85–90 % of affected fetuses.

2.5.3 Genetic Testing

Basically, genetic testing considering amniocentesis and chorionic villus sampling (CVS), the process of extracting amniotic fluid for analysis to determine the presence of genetic defects during pregnancy. This method is a confirmatory testing for chromosomal abnormalities detection which provides high accuracy as compared to previous described two methods. Amniocentesis is usually performed between 15 and 22 post-menstrual weeks of pregnancy. For earlier genetic

Fig. 2.6 Amniotic fluid
extraction using needle with
ultrasound guidance

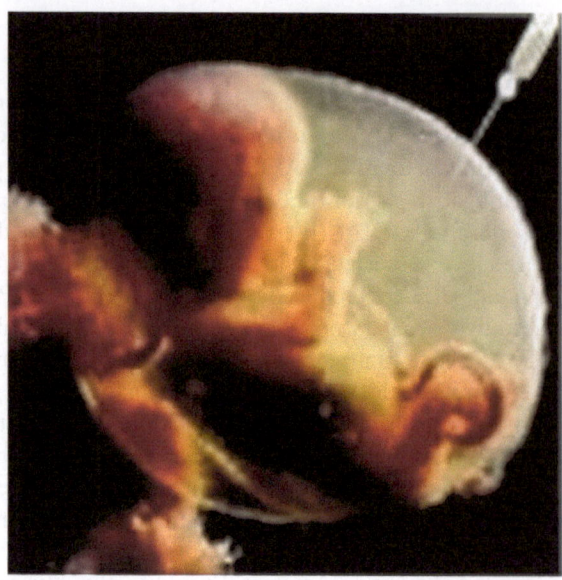

testing, CVS act as an alternative to the second trimester amniocentesis and can be performed from 10 weeks gestation age onwards. The procedure includes ultrasound guidance of a thin needle inserted through the abdominal wall to withdraw 2 tablespoons of amniotic fluid for analysis. Figure 2.6 illustrates the process of amniocentesis.

The risk for complications or miscarriage from having an amniocentesis performed is about 1 out of every 200 women, or 0.5 %. Complications include vaginal spotting or bleeding, leakage of amniotic fluid, severe cramping, fever, or infection. Meanwhile the risk of CVS during first trimester pregnancies is 1 in 100 women, or 1 %. According to FMF report, amniocentesis is also possible at 10 until 14 weeks of gestation. However, randomized studies have demonstrated that after early amniocentesis the rate of fetal loss is about 2 % higher (3 in 100 women).

2.5.4 Summary

Based on the literatures and consultation of collaborator hospital in Malaysia, we have summarized three different methods of Trisomy 21 detection, as shown in Table 2.1. Since genetic testing involves invasive examination with at least fetal loss probability of 1 in 100 women, if the fetuses have low chances of being Trisomies 21 babies, it is not recommended for pregnant women to perform these invasive examinations. Among the methods abovementioned, ultrasound prenatal screening for Trisomy 21 detection is the most favorable due to its intuitive,

Table 2.1 Method comparisons summary

Methods	Markers	Invasiveness	Gestation age	Risk (%)	Cost[a] (RM)	Operation durations	Recovery durations	Results availability
Ultrasound	Nuchal translucency, Nasal bone	non-invasive	1st Trimester	low	25	30 min	none	real time
Maternal serum markers	PAPP-A, free β-hCG, AFP	non-invasive	1st/2nd Trimester	low	350	<5 min	negligible	3 days
Genetic testing	Amniotic fluid, Chorionic villus	invasive	2nd Trimester	min 1–2 %	400 (QF-PCR) 580 (Karyotyping)	30 min	1 day	2 weeks

[a]Costs shown is applied on public hospita., it might be different at private or specialist hospital in Malaysia
[b]Quantitative fluorescent polymerase chain reaction (QF-PCR) is an alternative and rapid diagnostics method compare to full karyotyping

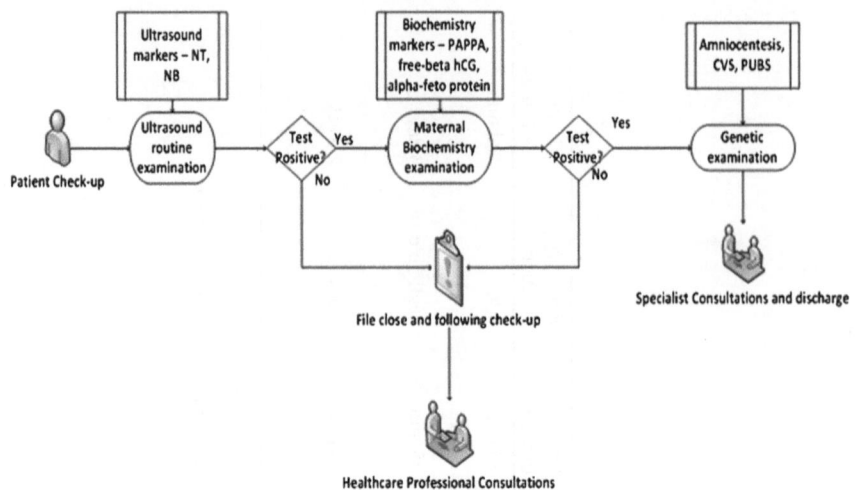

Fig. 2.7 Clinical work flow for Trisomy 21 detection process

Table 2.2 Detection rates for Trisomy 21 using single and hybrid combination methods

Methods	DR (%)
Maternal age (MA)	30
MA and maternal serum biochemistry at 15–18 weeks	50–70
MA and fetal nuchal translucency (NT) at 11–13 + 6 weeks	70–80
MA and fetal NT and maternal serum free b-hCG and PAPP-A at 11–13 + 6 weeks	85–90
MA and fetal NT and fetal nasal bone (NB) at 11–13 + 6 weeks	90
MA and fetal NT and NB and maternal serum free b-hCG and PAPP-A at 11–13 + 6 weeks	95
Amniocentesis or CVS	>99

non-invasiveness, flexibility, low risk, performance cost and time effectiveness. Figure 2.7 shows the process of clinical practices for chromosomal abnormalities detection, and Table 2.2 explained the detection rates of Trisomy 21 using each and hybrid combination methods.

With the aim of early detection in first trimester of pregnancies, ultrasound pre-natal screening at 11–13 weeks plus 6 days also appears as an advantage compared to biochemistry testing, or maternal serum markers assessment at second trimester pregnancies. Therefore, current practices in clinical field are using ultrasonic pre-natal examination, combining with maternal serum markers to assess the preliminary Trisomy 21 risk (Wee et al. 2010a; Nicolaides et al. 2008).

This book utilized the existing ultrasound imaging modality to improve the NT marker assessment in a semi-automated way while reducing human intervention. 3D volumetric images of NT are reconstructed for 3D boundary measurement rather than 2D weak echogenic lines.

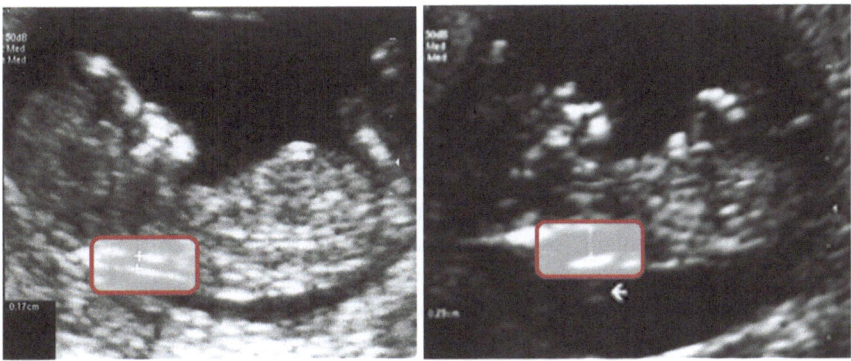

Fig. 2.8 Comparison of NT measurements with normal and high risk abnormal fetal: **b** Normal fetal 1.7 mm. **b** High risk abnormal fetal 2.9 mm

2.6 Two Dimensional Nuchal Translucency

In early 1990s, prenatal screening for high risk chromosomal defects such as Down's syndrome, triploidy, and Turner syndrome by using the combination of maternal age and NT thickness in the fetus within 11–13.6 weeks of gestation was introduced (Nicolaides et al. 1992, 1994). The term nuchal translucency (NT) was coined by Nicolaides et al. to describe the collection of fluid that is normally present behind the neck of the first trimester fetus. The stagnant fluids are obviously seen during 10–14 weeks gestation age, and then it will gradually decrease after 20 weeks and while making it difficult to detect the presence of key fold thickness. Fold thickness is a vital key marker to assess trisomy 21 of early pregnancy. 5 or 3.5 MHz abdominal ultrasound probe was used to scan the abdomen of pregnant women, where distance between fetal neck and the membrane cervical spine soft tissue in sagittal plane were measured as NT thickness. It should be essentially careful not to confuse the amniotic membrane as the layer of NT. According to FMF, NT measures considered abnormal were 3 mm and above and with 64 % sensitivity for trisomy 21 (Nicolaides et al. 1992). Figure 2.8 shows the example of normal and high risk abnormal ultrasonic NT thickness measurement.

2.6.1 Medicine Review and Related Researches

Meanwhile, some medical researchers claimed that fetus with large NT thickness (>3 mm) is associated with an increased risk for aneuploidy, congenital heart defect and other fetal anomalies (Hyett et al. 1995; Zosmer et al. 1999; Souka et al. 2001). Some publications also indicate that an increased NT thickness that is more than 2.5 mm in between 10 and 13 weeks plus 6 days has also been associated with an increased risk of congenital heart and genetic syndrome (Pandya et al. 1994). Based on the past research in Harris Birthright research center for fetal medicine (Snijders et al. 1998), have coordinated the largest research to assess NT accuracy. It was conducted at 22 ultrasound centers in England on 96,127 women who were 10–14 weeks pregnant. The risk for trisomy 21 was calculated by multiplying the NT probability ratio by the prevalence of this trisomy at different maternal and gestational ages. Findings shows that 326 cases are found to be Trisomy 21 and among them, 231 cases or 71.2 % are found NT thickness are higher than 95th percentile. This has been proved that NT is the powerful marker for trisomy 21 screening. In 2004, the populations examined shows the definition of the minimum abnormal NT thickness is ranged from 2 to 10 mm (Nicholaides et al.). Prospective studies in more than 2,00,000 pregnancies which includes more than 900 Trisomy fetuses, indicates that NT screening assist in Trisomy detection for more than 75, with false positive rate of 5 % (FMF report). The importance of measuring NT as a screening tool can be evaluated from the fact that all over Europe, America and UK, NT measurement is included in their prenatal screening programs.

Hereafter the measurements of NT appear as a powerful screening marker during first trimester of pregnancy—early detection of abnormal findings in pregnancy is pivotal in establishing premature evaluation of chromosomes and possible structural defects on a targeted basis (Lee and Kim 2006). Trans-abdominal sonographic examinations are widely used to show the mid-sagittal image of the fetal neck to measure the nuchal fold. Trans-vaginal ultrasound of NT appears to be a more accurate method (Braithwaite and Economides 1995) due to increased resolution. Conversely, when using a trans-abdominal probe, the examiner possesses a wider range of maneuverability to obtain the correct mid-sagittal view of the fetus (Cullen et al. 1989). Fetal NT can be measured successfully by trans-abdominal ultrasound examination in about 95 % of cases; the rest cases are necessary to perform trans-vaginal sonography (FMF report). With the fact that NT thickness is known to be reliable marker for Trisomy 21 assessment, the ability to achieve a convinced measurement using manual B-mode ultrasound caliper is dependent on appropriate training and adherence to a standard technique in order to achieve uniformity of results among different operators.

Hence, fetal Medicine Foundation has promoted standardization in the assessment of NT which should follow the following criteria (FMF report); (a) Gestation should be between 11 and 13 +6 weeks, (b) Image is magnified so that head and upper thorax are included in the screen, (c) A mid sagittal view of fetal profile is obtained with ultrasound transducer being held parallel to longitudinal axis of nuchal translucency, (d) Crown hip length to be within the 45–84 mm, (e) Fetus must be in neutral position, as hyperextended or flexed neck will results 0.6 and 0.4 mm deviation respectively. (f) More than one measurement must be taken and maximum thickness in true sagittal plane is considered as NT thickness, (g) The calipers should place on lines that define NT thickness precisely; crossbar of caliper should be such that is hardly visible as it merges with the white line of the boundaries and not in the nuchal fluid. This is the most difficult requirement to realize under control. Generally, the

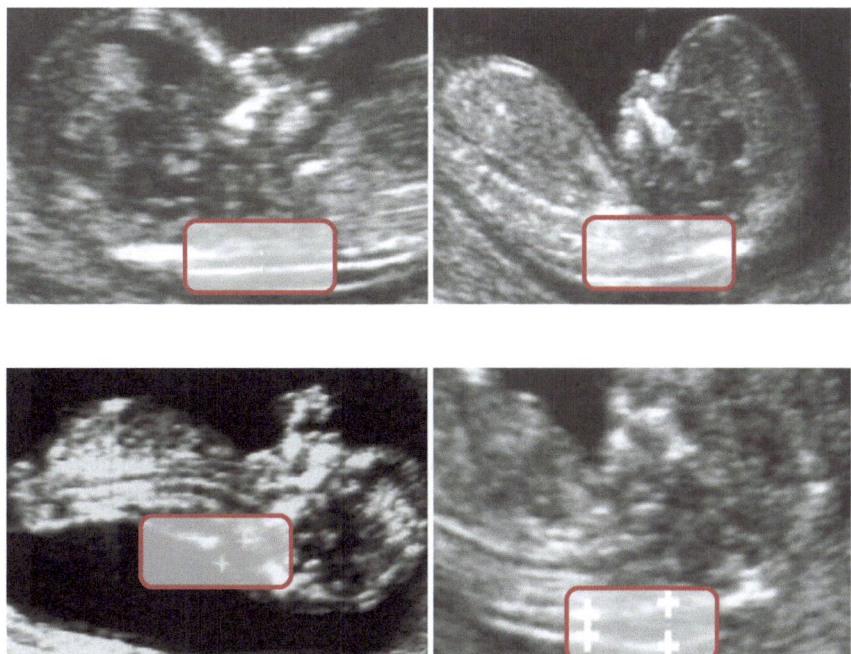

Fig. 2.9 Ultrasonic marker measurements; nuchal translucency at 12 weeks fetuses. **a** Correct measurement adhere to FMF protocol. **b** Hyperextended neck. **c** Flexed neck. **d** Maximum measurement of NT should be taken

measurement of fetal NT layer takes at least 15 min (Taipale et al. 1997), if the fetus is not in the right position, the overall prenatal screening will consume longer time. Figure 2.9 shows some example of NT measurement in different position and condition.

However, review of images by an experienced operator indicated that assessment may have been hampered either by poor magnification and unfavorable section or by untrained operator. As with screening based on NT, currently it is imperative that sonographers who undertake risk assessment by examination of fetal profile receive appropriate training and certification of their competence in performing such a scan. Reproducibility studies suggest that reproducibility of measurement is variable among groups and poor in some studies (Kanellopoulos et al. 2003; Bekker et al. 2004; Malone et al. 2004). It is possible that learning curve for this measurement is much longer for NT measurement (Cicero et al. 2003).

In some observational studies (Roberts et al. 1995; Kornman et al. 1996), the scans were often carried out at inappropriate protocol and the sonographers were either not trained adequately or they were not sufficiently motivated to measure NT. These methodological problems are further highlighted by a research of

Table 2.3 Correlation between nuchal translucency thickness, chromosomal abnormalities and alive well percentage

Nuchal translucency (mm)	Chromosomal abnormalities (%)	Alive well (%)
<95th centile	0.2	97
95–99th centiles	3.7	93
3.5–4.4	21.1	70
4.5–5.4	33.3	50
5.5–6.4	50.5	30
≥6.5	64.5	15

47,053 singleton pregnancies examined at 6–16 weeks (Wald et al. 2003). In 23 % of the patients no valid NT measurement was taken because the scans were carried out at inappropriate gestations or the sonographers were unable to obtain a measurement or none of the images were deemed to be of an acceptable quality.

2.6.1.1 Existing Limitation

From the medical researches earlier mention, it is known that trisomy 21 characteristic can be extracted from fetal ultrasound image with measured nuchal translucency thickness. Previous researches have concluded that minor inaccuracies in NT measurement as small as 25 % or 0.5 mm will have very significant negative impacts upon abnormality detection, reducing detection rates by 18 % (Moratalla et al. 2010; Abele et al. 2010). Table 2.3 summarized the correlation of nuchal translucency thickness with chromosomal abnormalities and alive well percentages.

Although FMF have developed a standardize protocol for NT assessment in 11 until 13 weeks + 6 days, it is recognized hardly to implement and practices in clinical implementation and realization. Inter and intra observer variability using conventional B-mode ultrasonic marker measurement (Pandya et al. 1995; Kanellopoulos et al. 2003) is still unavoidable. Accurate calipers placement on 2D echogenic lines boundaries are certainly a challenging problem, as shown in Fig. 2.10, therefore, the consistency of measurement cannot be guaranteed and always subject to human errors, technical difficulties, patient loads, and longer time consumptions.

2.6.2 Engineering Review and Related Researches

The application of ultrasound imaging to detect fetal abnormalities in early pregnancy has aroused great attention of genetic workers. There are also some related researches done in engineering field. Much attention now is focused on techniques to segment and measure NT marker only in 2D ultrasonic images. Efforts

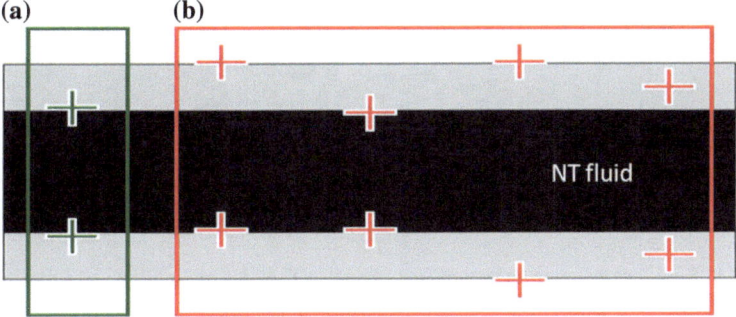

Fig. 2.10 Displacement of ultrasonic calipers for NT thickness boundary measurement. **a** Correct calipers placement. **b** Incorrect calipers placement

have been made by several investigators worldwide to try to find an approach for automation NT in two key procedures; first is the automatic distal and proximal echogenic lines detection and second is to measure the NT marker thickness in automated way, in order to reduce amount of human intervention. None of the 3D approaches dedicated for NT marker were found. In fact, there are very few papers dedicated on ultrasound imaging reporting automatic or semi-automatic NT measurement up to now. It reveals the fact that ultrasound fetal images are the difficult data to deal with, and therefore, the problem is still far from being solved until now.

The first scientific paper working on NT automation is Bernardino et al. 1998. They proposed simple image processing technique; histogram equalization for contrast enhancement and Sobel edge operator; to extract the upper and bottom echogenic lines of NT marker. The Sobel operator implements two 3×3 kernels which are convolved with the sources image; $A(i,j)$ to calculate approximations of the derivatives—one for horizontal changes $h_x(i,j)$, and another for vertical $h_y(i,j)$ as shown below;

$$G\,(i,j) = \sqrt{\left(h_x^2\,(i,j) + h_y^2(i,j)\right)} \qquad (2.5)$$

$$\theta(i,j)\,\text{arctanget}\left(\frac{h_x^2\,(i,j)}{h_y^2(i,j)}\right) \qquad (2.6)$$

$$h_x = \begin{bmatrix} -1 & 0 & 1 \\ -2 & 0 & 2 \\ -1 & 0 & 1 \end{bmatrix} \;;\; h_y = \begin{bmatrix} 1 & 2 & 1 \\ 0 & 0 & 0 \\ -1 & -2 & -1 \end{bmatrix} \qquad (2.7)$$

where, $G(i, j)$ is the gradient magnitude, $\theta(i, j)$ is the gradient phase

This simple Sobel operator using a threshold specified by the user on the magnitude of the gradient for detection a variable number of image edges. But problem arise with no single image features can provide reliable NT boundaries for thickness measurement. The location of the edge is entirely determined by local

(a)   (b)   (c)

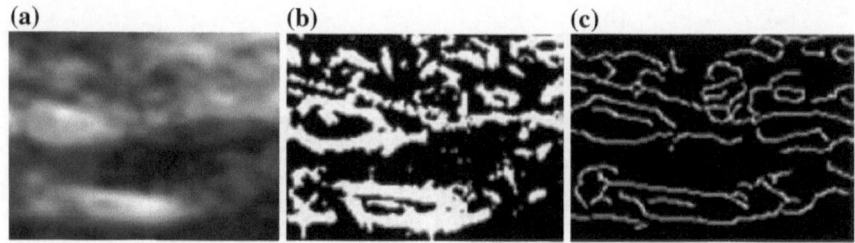

Fig. 2.11 Resultant of edge detection. **a** Original NT region of interest. **b** Sobel operator. **c** Canny operator

evaluation of single image feature such as the intensity or the intensity gradient. It is therefore impossible to detect the border of NT layer correctly in single image feature. Figure 2.11 illustrates the difficulty of simple edge operator implementing on ultrasonic images.

Conventional edge detection such as Sobel and Canny techniques has a drawback in NT measurement, as more than two echogenic lines will be mapped within the output image. A decade later, a method for semi-automated NT border measurements based on dynamic programming was proposed by Lee et al. in (2007). They presented a computerized method of detecting the border of NT layer by minimizing a cost function using dynamic programming. Thanks to the matured development of ultrasound speckle noise filter; nonlinear anisotropic diffusion techniques are implemented as their pre-processing before NT edge segmentation. The anisotropic diffusion filter having good characteristics to preserve NT image edge features while blurring the area inside the NT layer. The general diffusion equation proposed by Perona and Malik (1990) is as follows;

$$\frac{\partial I}{\partial t} = div\left(c\left(x, y, t\right) \nabla I\right) = \nabla c \cdot \nabla I + c\left(x, y, t\right) \Delta I \qquad (2.8)$$

with the condition,

$$I\left(t = 0\right) \equiv I_0$$

where ∇ is the gradient operator, div is the divergence operator, $c(x, y)$ is the diffusion coefficient, t is the diffusion time, Δ is the Laplacian of I, and I_0 is the initial image. $C\left(t\right)$ controls the rate of diffusion and is usually chosen as a function of the image gradient so as to preserve edges in the image. Perona and Malik proposed C (model) has the following two forms:

$$c(|\nabla I|) = \exp\left[-\left(\frac{|\nabla I|}{k}\right)^2\right] \qquad (2.9)$$

$$c(|\nabla I|) = \frac{1}{1 + \left(\frac{|\nabla I|}{k}\right)^2} \qquad (2.10)$$

where k is the edge magnitude factor, correlated to the contradiction degree's balance of edge preservation and smoothing factor, and final smoothing outcomes are influenced by diffusion time t. The basic idea of P-M model is using $c(|\nabla I|)$ to control the diffuse proliferation on the initial image. The model achieving adaptive diffusion based on image gradient magnitude. At the edges with large gradient modulus, $c(|\nabla I|)$whichever is less; the model is weak in smooth implementation to protect the edge information. In the homogenous areas gradient modulus is smaller, $c(|\nabla I|)$become larger; the model has more smoothing effect. Adaptive selection of smoothing degree at the edge and homogeneous region can help to identify the boundary location, and solved the contradiction between de-noising and edge retention.

For the NT layer segmentation, cost functions are built for each of the borders of the NT layer. Let's assumed all the possible borders B_n can be considered as polylines with n nodes;

$$B_n = \{P_1, P_2, P_3 \ldots, \ldots P_n\} \tag{2.11}$$

where $P_1, P_2, P_3 \ldots$ are the neighbouring pixels in x axis; n is the number of contours lines in horizontal length. Figure 2.12 illustrates the zone discrimination for NT image features and line border definition of B_1 and B_2.

The minimized cost function can be expressed as a sum of local costs along a candidate border B_N;

$$C(B_N) = C_f(P_1) = \sum_{i=2}^{N} (c_f(P_i) + C_g(P_{i-1}, P_i)) \tag{2.12}$$

When point $P_{i>1}$, local cost function terms $C_f(P_i)$ and $C_g(P_{i-1}, P_i)$ are defined as follows;

$$C_f(P_i) = \sum_{j=1}^{k} w_j f_j(P_i) \qquad i = 1, \ldots, N \tag{2.13}$$

$$C_g(P_{i-1}, P_i) = W_{k+1} g(P_{i-1}, P_i) \tag{2.14}$$

$$g(P_{i-1}, P_i) = \left| d(P_i) - d(P_{i-1}) \right|^2 \quad i = 2, \ldots, N \tag{2.15}$$

By combining Eqs. 2.13, 2.14 and 2.15, minimized cost function is;

$$C(B_N) = C_f(P_1) + \sum_{i=2}^{N} \left(\sum_{j=1}^{k} w_j f_j(P_i) + w_{k+1} \left| d(P_i) - d(P_{i-1}) \right|^2 \right) \tag{2.16}$$

Where $f_j(P_i)$ are image feature terms, w_j is a weighting factor, $k = 3$ equal to the number of image features been considered, $|d(P_i) - d(P_{i-1})|^2$ is geometrical force term and d is the vertical distance between the border being estimated and a

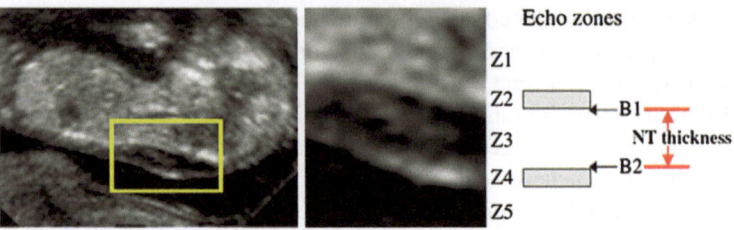

Fig. 2.12 NT border definitions and echo zones (Lee et al. 2007)

reference node. Their weighting factors are determined empirically for each border with relative constraint as follows;

$$|w_1| + |w_2| + |w_3| + |w_4| = 1 \tag{2.17}$$

According to the NT image features characteristics, referring to Fig. 2.12, the border B1 is below the bright region and above the dark region, meanwhile, border B2 is above the bright region and below the dark region. In other words, the characteristics of upper border B1 is opposite the lower border B2, therefore, weighting factors of image feature terms have to be opposite signs respectively. By considering this image feature characteristics, cost function $C(B_N)$ for B2 will be calculated in advanced by taking the horizontal line as reference line for $g(P_{i-1}, P_i)$.

Based on Eq. 2.16, the first local cost term $f_1(P_i)$ measures the mean intensity of 3 pixels below a pixel P_i in order to detect a pixel above echo zone Z4. Second term $f_2(P_i)$ calculated the mean intensity of two pixels above a pixel P_i in order to detect a pixel below dark NT region at zone Z3. The third cost term $f_3(P_i)$ computed the downward intensity gradient at upper edge of echo zone Z4 by using vertical gradient operator $[1 \ 0 \ -1]^T$. For the final cost function term $g(P_{i-1}, P_i)$ or $|d(P_i) - d(P_{i-1})|^2$, it is the vertical distance between the estimated border and a reference line for border continuity concern. For their proposed technique, B2 is calculated before B1; therefore, the estimating cost term $g(P_{i-1}, P_i)$ in B1 control will take B2 as its reference line (Fig. 2.13).

The reason for Lee et al. (2007) to implement the DP back-tracking technique is to avoid local minima which could cause missed tracking of the optimum minimized cost function calculated previously. DP is commonly applied to optimize problems, in our case: tracing minimum cost term globally within the ROI in backward propagation horizontal polylines.

Although Lee's method improves the NT border continuity by local cost term $g(P_{i-1}, P_i)$ and reduces the problem of operator variability using manual NT tracking and measurement, however, it remains several existing limitations; firstly, it does not solve the difficulty of true mid-sagittal plane selection which coincides with NT marker with maximum thickness: the plane selection work scope is pre-stage of their research data experimental simulation. Therefore, results of their semi-automated may underestimate the NT thickness; Second limitation is the choice of orientation for foetus position; their proposed technique can only

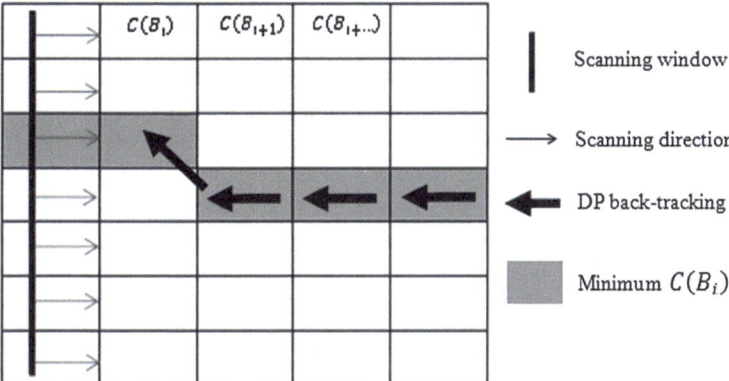

Fig. 2.13 Border line B2 detection using cost function $C(B_N)$ computation followed by dynamic programming (DP) back-tracking

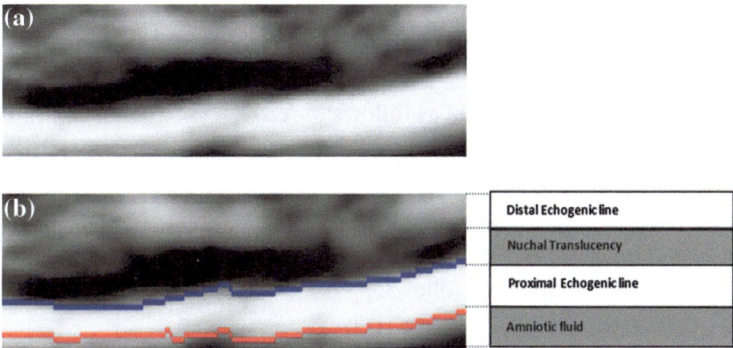

Fig. 2.14 Limitation of incorrect initiation line tracking. **a** Original NT ROI. **b** Missed tracked NT at proximal echogenic line

be applied on horizontal foetus position, which limits the method feasibility: the cost function $C(B_i)$ considers only horizontal neighbouring B_l node, therefore, DP back-tracking for global minimizes cost function in vertical NT position is not realizable.

Next, the manual cropping of NT region of interest (ROI) influences the automatic echogenic lines tracing heavily, their method is limit with fine ROI cropping excluding placenta beneath bottom NT line wisely, which is not the easy case in most of the ultrasonic images. Enlarged manual ROI cropping will change their weight factor characteristics as it refers to Fig. 2.12, simultaneously, final computation values of cost function for each pixel P_i can be changed. Hence, their NT layer are tracked one after another, by taking the first tracked line as reference line, if the initial first line was tracked at wrong position: bottom of proximal lines, consequently, second NT line will be tracked perfectly in wrong position. Figure 2.14 illustrates this limitation of incorrect initiated line and this leads to miss tracked NT layer.

Their proposed cost function calculation is heavily dependent on their weight factors, refer to Eq. 2.17, and the weights are not necessarily the same for both NT layers. Therefore, the choice of the cost function appears based on empiric consideration and it may further influent the quality of results explicitly.

Last but not least, the cost term $g(P_{i-1},P_i)$ for B2 distance calculation at pixel P_i from a reference line, which assumed as straight horizontal lines will cancel each other eventually. It only shows its effect when calculating cost term for B1, where previous tracked B2 will be taken as its reference line.

Due to this limitation, we have also proposed an iterative algorithm for both echogenic line borders simultaneous detection on 2D B-mode ultrasound images. The prior step of the proposed technique is the same with previous publication, which requires manual ROI containing NT layer thickness selection. Let's assume the acquired ROI is an $M \times N$ rectangle, and then all possible borders T_N are considered as polylines with N nodes:

$$T_n = \left[P_1, P_2, P_3, \ldots, P_{n-1}, P_n\right] \tag{2.18}$$

Where the pixels P_{n-1} and P_n are horizontal neighbors and n is the horizontal length of a contour line. The function of NT backbone $B(\Upsilon)$ is build according to reference point r, which is defined as follows:

$$\Upsilon_{1,2} = min\left[f(P_{1,n})\right] \tag{2.19}$$

The term $f(P_{1,n})$ measures the intensity gradient and intensity of pixels along P_1 and P_n, as shown in Fig. 2.15. Applied Eq. 2.19, the $B(\Upsilon)$ is formulated based on linear equation, as expressed follows:

$$y_j = \nabla B(\Upsilon)x_i + \Upsilon_1$$
$$i = 1,\ldots n \quad j = \Upsilon_1,\ldots,\Upsilon_2 \tag{2.20}$$

$$\nabla B(\Upsilon) = \frac{|\Upsilon_1 - \Upsilon_2|}{n} \tag{2.21}$$

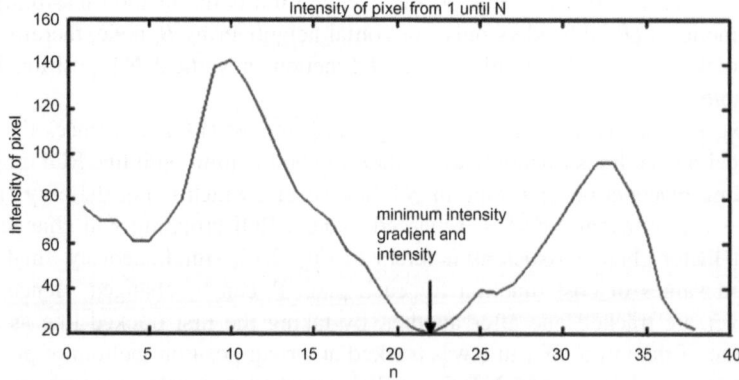

Fig. 2.15   Intensity gradients and intensity of pixels along P_1 and P_n

Fig. 2.16 Formation of NT backbone $B(\Upsilon)$ using both reference points Υ_1 and Υ_2

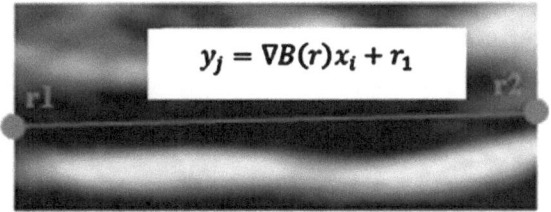

$$y_j = \nabla B(r)x_i + r_1$$

where x_i and y_j are the coordinate along this linear equation, Fig. 2.16 illustrates the linear equation coincide with both reference points Υ_1 and Υ_2.

The bidirectional forward propagation tracking process is used to scan through the NT edges of upper and lower boundaries within the $M \times N$ ROI referring to $B(\Upsilon)$, and stored in the array of T_{N1}, T_{N2}, as shown below;

$$T_{N1} = max\left[\nabla ROI\left(x_i, y_j - d_{1i}\right)\right] \tag{2.22}$$

$$T_{N2} = max\left[\nabla ROI\left(x_i, y_j - d_{2i}\right)\right] \tag{2.23}$$

where d_{1i} and d_{2i} are y-coordinates for maximum intensity gradient of both upper and lower border, the NT thickness was taken along every five pixels of polylines T_{N1} and T_{N2}. The maximum thickness of the subcutaneous translucency between skin and the soft tissue overlying the cervical spine should be measured. Therefore, the largest thickness is recorded as the NT measurement and calibrated with scale of ultrasound image to get the exact thickness in millimeter, as shown in Figs. 2.17, 2.18, 2.19.

Catanzariti et al. (2009) have also proposed an improved cost function for NT border segmentation. They modify the cost function in Eq. 2.12 by removing the weight factors, which indicative through empirical consideration. They believe this can help enhancing the automation process of NT layer measurement. Same to all previous semi-automated techniques, their proposed algorithm needs a manual NT ROI identification before introducing to their improved cost functions, as follows; Considering a NT ROI with dimensions $N \times M$, borders of polylines are;

$$B_N = \{P_1, P_2, P_3, \ldots, P_{N-1}, P_N,\} \tag{2.24}$$

where P_{N-1} and P_N are the adjacent pixels in horizontal; N equals to length of estimated border. Referring to Eq. 2.12, they have separated the cost function to two different functions for each upper and lower border respectively. The cost function to minimize lower border is as follows;

$$C_l(B_n) = C_l(B_{n-1}) + \left\{-\frac{\partial f}{\partial y}(P_n) + Z_l(P_n) + f_{\text{adj}}(P_n, P_{n-1})\right\} \tag{2.25}$$

Fig. 2.17 Experimental result of maximum thickness NT measurement. **a** Sample original image. **b** Edge tracking and NT measurement

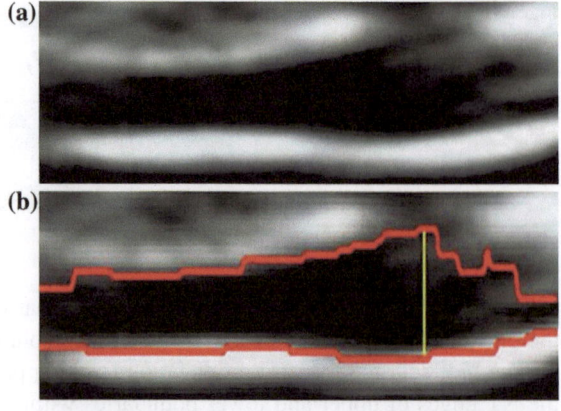

Fig. 2.18 Comparison of various edge detectors. **a** Original image. **b** Sobel detector. **c** Canny detector. **d** BIFP edge racing

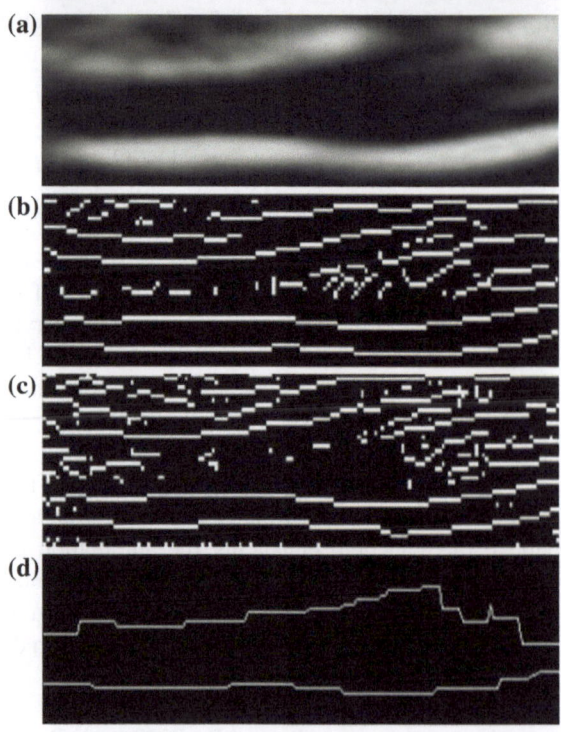

where at node $n = 1$

$$C_l(B_1) = -\frac{\partial f}{\partial y}(P_1) + Z_l(x, y) \tag{2.26}$$

$$Z_l(p) = \begin{cases} 0 \ if \ \frac{\partial^2 f}{\partial \theta^2}(p) \times \frac{\partial^2 f}{\partial \theta^2}(t) < 0 \\ \varsigma \ otherwise \end{cases} \tag{2.27}$$

$$f_{adj}(P_n, P_{n-1}) = \begin{cases} 0 \text{ if } d(P_n, P_{n-1}) \leq 1 \\ \varsigma \text{ otherwise} \end{cases} \quad (2.28)$$

The first local cost term $\frac{\partial f}{\partial y}(P_n)$ aims to replace both $f_1(P_i)$ and $f_2(P_i)$ at Eq. 2.13. It consists of the image derivative along vertical direction to consider the energy deriving from the image features as edges or lines. The second cost term is a second order derivatives computed along gradient direction, as a replacement for gradient operator $[1 \ 0 \ -1]^T$ proposed by Lee et al. (2007). Lastly, f_{adj} is built to replace $g(P_{i-1}, P_i)$ term, to enforce borders continuity by penalising consecutive pixels distance larger than one pixel. Similar to Eq. 2.25, upper border cost function is built as follows;

$$C_u(B_n) = C_l(B_{n-1}) + \left\{ \frac{\partial f}{\partial y}(P_n) + Z_u(P_n) f_{adj}(P_n, P_{n-1}) + f_{pos}(B_n, B_n) \right\} \quad (2.29)$$

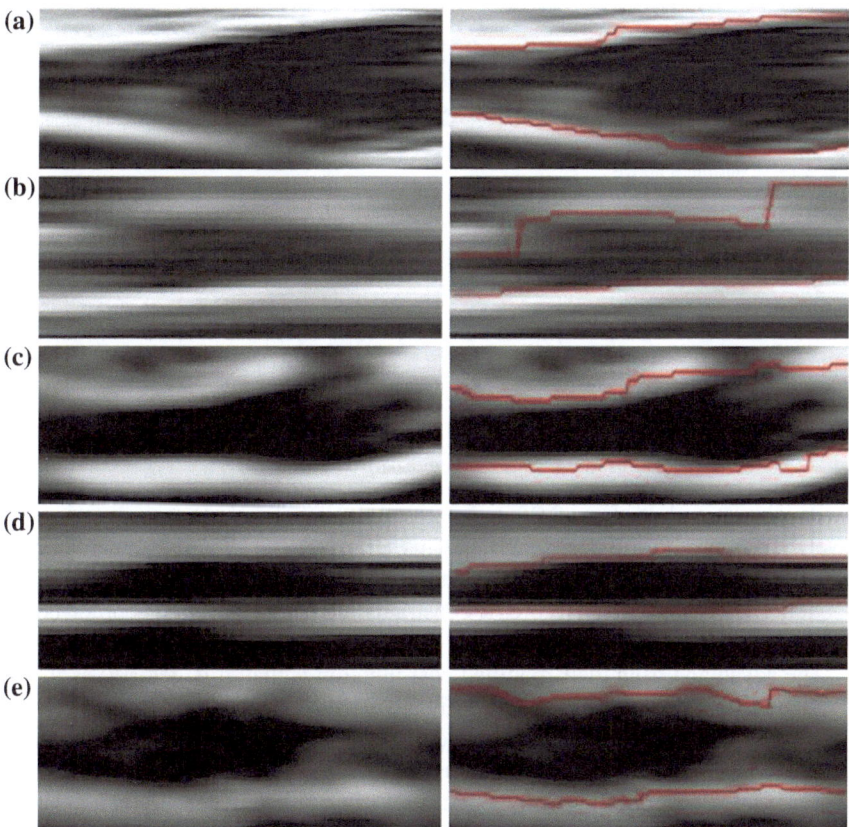

Fig. 2.19 Experimental results on 2D B-mode Ultrasonic NT images, left are original sample images, right are the findings of BIFP algorithm

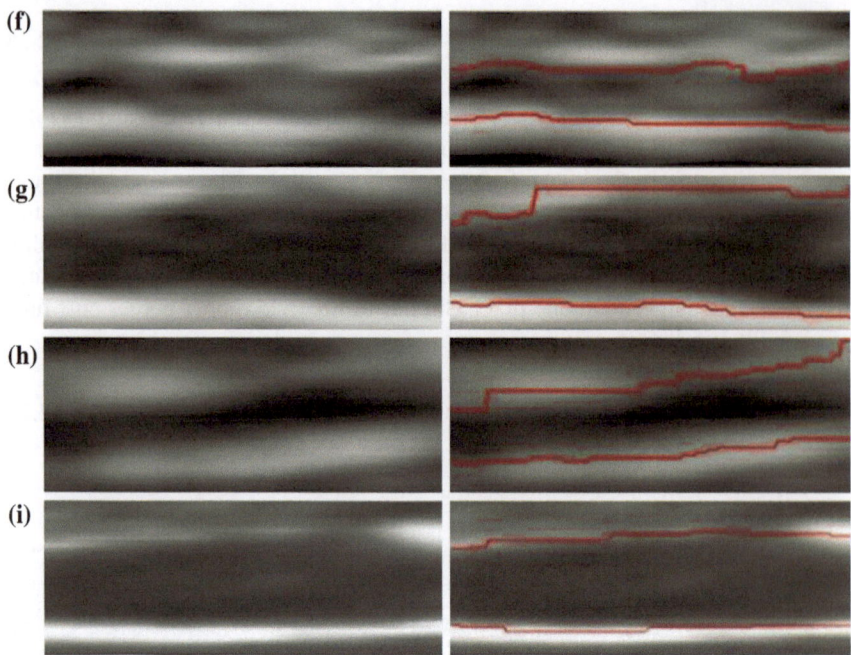

Fig. 2.19 Continued above

where at node $n = 1$

$$C_u(B_1) + \left\{ \frac{\partial f}{\partial y}(P_1) + Z_u(x, y) + f_{\text{pos}}(B_1, P_1) \right\} \tag{2.30}$$

$$Z_u(p) = \left\{ 0 \text{ if } \frac{\partial^2 f}{\partial \theta^2}(p) \times \frac{\partial^2 f}{\partial \theta^2}(q) < 0 \atop \varsigma \text{ otherwise} \right\} \tag{2.31}$$

By observing both Eqs. 2.25 and 2.29, a new local term is introduced exclusively for upper border cost function. This term consists of a sigmoidal function which weighs the relative pixels distance between upper border and its corresponding lower border, therefore, it constrains the estimated upper border is stay above the lower NT border.

The advantage from their modified cost function is that, it is general; as it does not depend on weight tuning for each image. Nevertheless, it is still inherent with others existing limitation from Lee's method includes true sagittal plane selection; enlarged ROI manual cropping and fetus position.

Since late 2010, the first commercial tool of semi-automated 2D NT measurement system is called SonoNT^{TM} using GE Voluson 730 Expert (RAB 4-8L probe, Milwaukee, WI, USA) was reported. Moratalla et al. (2010) and Abele et al. (2010) have implemented this new commercial tool to investigate the operator inter and intra variability, and compared to conventional NT manual measurement method.

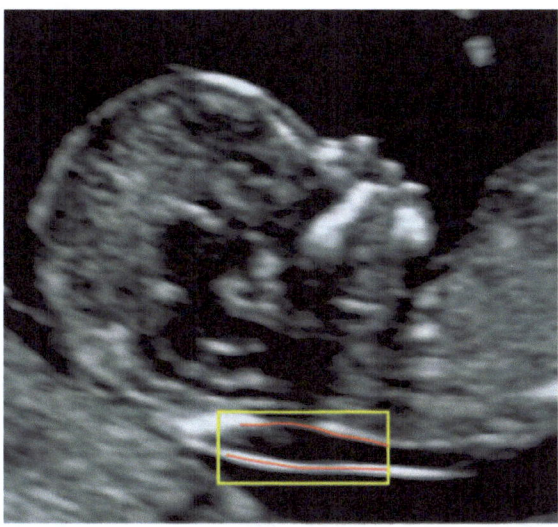

Fig. 2.20   Semi-automated process of SonoNT^TM, box placed by operator and the lines draw by system automatically (Moratalla et al. 2010)

An adjustable box has to be placed on relevant area at the back of the fetal neck, where the semi-automated system will interrogates the whole length of nuchal membrane within the marked box and draws the edge lines through the center of nuchal membrane and soft tissue overlying cervical spines respectively. This system utilizes the gradient and brightness information inside the box to define their drawing lines. Unfortunately, due to its commercial copyright and patents, details of their algorithms are not reported. Figure 2.20 illustrates the process of SonoNT^TM to measure NT thickness. In order to calculate the vertical distance of NT thickness, each point on one line is virtually connected to all possible points on the other line, the final NT length is the longest among all the minimum distances between the lines.

The vital key difference of this new commercial NT tool as compared to the previous gold standard developed by FMF; the automated drawing line that defines the edge of NT layer are laid on the centre of nuchal membrane rather than at its inner border. This is due to their system requirement to magnify the ultrasound fetal images (either pre- or post-freeze zoom) which results in thickening of the lines that defines fetal NT. Consequently, the translucent area between NT layers become smaller and may lead to measurement underestimation. This phenomenon was not reported in the studies since early 1990s. It is believed that pre-processing and post-processing imaging by latest GE US system which it includes speckle reduction imaging (SRI) and harmonics application contribute to the factor of thickening membrane lines (Moratalla et al. 2010). Therefore, the GE healthcare technology segments the NT border at the point of maximum echogenicity which normally lies in the centre of membrane lines. This is different with the measurement protocol developed by FMF Fig. 2.21.

Issue on relying SonoNT^TM to measure NT thickness and replace expertise have arisen an argument; blasphemy or oblation to quality, as reported by Ville (2010). It is exposed to two main types of drift from the expected clinical and

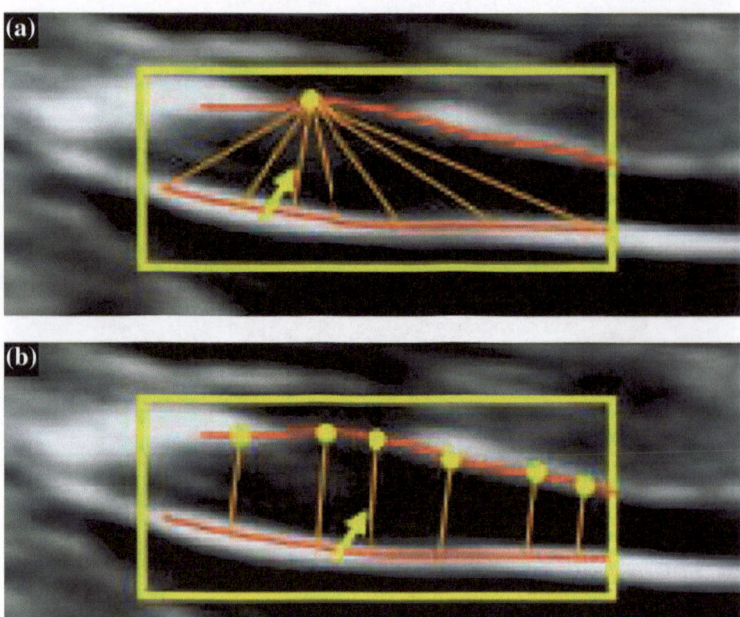

Fig. 2.21 Nuchal translucency thickness measurement. **a** Virtual points connection for minimum distance calculation. **b** Longest length among minimum distance lines is considered as NT thickness (Moratalla et al. 2010)

economics benefits; misuse and abuse of automation. Abuse occurs when the design function does not fit the clinical expectation, and misuse arises when operator overreliance on automation and their role becomes by-product of automation. Crude errors in NT measurements are generated by this semi-automated system if the selected ROI box encompasses more of nuchal area; this is the same typical error for all previous described semi-automated NT researches (Lee et al. 2007; Catanzariti et al. 2009; Bernardino et al. 1998). Figure 2.22 illustrates the example of semi-automated measurement error.

Furthermore, one argues that this semi-automation system is not useful for well-trained operators, as each individual interprets results differently. Yet, it is still questionable to accept this new method when there is no published references range, as compared to the FMF method which establishes through cumulative studies over the last 10 years.

2.6.2.1 Summary

Among all previous engineering work done, dedicated on automatic or semi-automatic fetal measurements, research topics are focused on nuchal translucency thickness segmentation and measurement using 2D ultrasonic images.

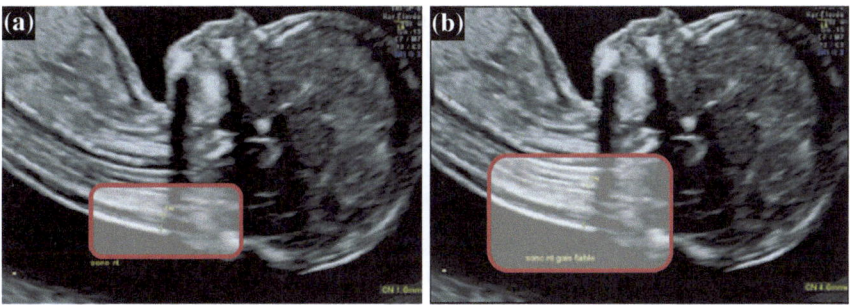

Fig. 2.22 False reading when larger ROI box placed over nuchal area. **a** Correct NT measurement 1.6 mm with fit ROI box. **b** Abuse NT measurement 4.6 mm with larger ROI box. (Ville 2010)

However, manipulation of true mid-sagittal plane selection was never resolved in which causing underestimation of NT thickness, while the position of fetus is limited to horizontal position and limitation of manual NT ROI cropping excluding non NT layer region.

These reveal that the existing methodological problems needs a 3D computing for NT measurement, in which measured 3D NT structure appears explicit assembly as compared to 2D inherent lines boundary. The current sonographer picks the three best 2D sagittal planes upon on their experiences using hand-eye coordination, and average of three NT measurements is taken as the final marker thickness. This inefficiency can be resolved within 3D NT reconstruction and incorrect plane selection can be avoided. Besides, with existing 2D images only the recorded images can be reviewed, even minor changes cannot be made. With 3D NT approaches, the actual viewing planes can be manipulated using the reconstructed saved volume data; the NT measurement can be re-calculated to aid reassessments and validation.

Our efforts are to find an approach for boundary detection in ultrasonic NT images which is less reliant on human operators. As it reduces the amount of human intervention, it will also reduce inter-observer variability and the intra-observer variability is expected to be reduced; consequently, drifting problem in measurements over time in longitudinal studies is reduced. Therefore, we have extended the current NT measurement from 2D ultrasonic marker to 3D volumetric ultrasound in order to overcome all the limitations above.

2.7 Three Dimensional Ultrasound Applications

With continuous improvement of ultrasound equipment and innovative technology in current research and development, three-dimensional (3D) ultrasound technology has been used in clinical research and diagnosis, particularly in prenatal care aspects. In late 1980's, 3D ultrasound imaging becomes reality due to the rapid development of computing technology in terms of both hardware and software. There are three different 3D ultrasound imaging visualization, which includes surface rendering, transparent volumetric rendering and multi-planar reformatting (MPR). Nevertheless, the visualization of these 3D ultrasound imaging are still heavily influenced by quality of the 2D image. In the early 1990s, 3D images of first trimester pregnancies were presented (Bonilla-Musoles et al. 1995; Kelly et al. 1992).

3D medical images have been found to be more valuable and powerful fetal diagnostic tool (Fenster et al. 2001). The visualization of conventional 2D medical data is rather trivial while visualization of 3D volumetric data is not (Wee et al. 2011). The major application of 3D ultrasound in current prenatal screening tends to inspect qualitative physical abnormalities and quantitative masked volume measurement. Qualitatively, much of these work have been concerned with the detection of fetal physical abnormalities (Baba et al. 1999), for instance: facial, cleft lips, limb and other physical anatomy development (Lee et al. 1995). In 1995, Nelson and Pretorius have been using 3D ultrasound imaging to evaluate skeletal dysplasia, abnormalities leading to a small thorax and neural tube defects (NTD). NTD is a failure developing fetal spine which do not close properly, resulting anencephaly and spina bifida.

Quantitatively, the 3D measurements are mostly focused on masked volume estimation of placental, fetal and gestational sac (Blaas et al. 1998; Hafner et al. 1998). The estimated ROI volumes are either masked manually for multiple, individual slices or by masking a structure that has been isolated using an editing tool. Previous researches have reported 3D ultrasound imaging for volume measurement of fetal lumbar spine (Schild et al. 1999), fetal volume and weight estimation (Rankin et al. 1993), fetal liver (Laudy et al. 1998), and fetal lung (Pohls and Rempen 1998). Another concern of obstetrics volume measurement is placental volume around mid-pregnancy for birth weight estimation (Howe et al. 1994). Besides, the usages of trans-vaginal 3D ultrasound have been reported to investigate embryos shape and volume during early pregnancy (Blaas et al. 1998). However, in Malaysia, 3D ultrasound scanning is not covered in the routine prenatal screening protocol.

In addition to obstetrics and gynecology application, the usefulness of 3D ultrasound imaging has also reported to cover range from neurology (Rankin et al. 1993) for tumors diagnostics during brain surgery; neonatal ventricular volume measurement (Nagdyman et al. 1999; Kampmann et al. 1998) to cardiology (Martin et al. 1990; Arbeille et al. 2000; Gopal et al. 1997; Ofili and Nanda 1994; Salustri and Roelandt 1995; Magni et al. 1996). Some reports show high feasibility in 3D volume visualization and measurement for accurate atherosclerotic plaques diagnostics (Fenster and Downey et al. 1996; Rosenfield et al. 1992; Allott et al. 1999), prostate gland volume measurement (Aarnink et al. 1995; Basset et al. 1991; Nathan et al. 1996; Terris and Stamey 1991), breast imaging (Fenster and Downey 1996; Fenster et al. 1995), and ophthalmology (Downey et al. 1996).

Among all the previous literatures, 3D ultrasound techniques are not widely applied on NT application. Clinical personnel follow the gold standard developed by FMF for B-mode ultrasonic NT marker measurement despite it contains

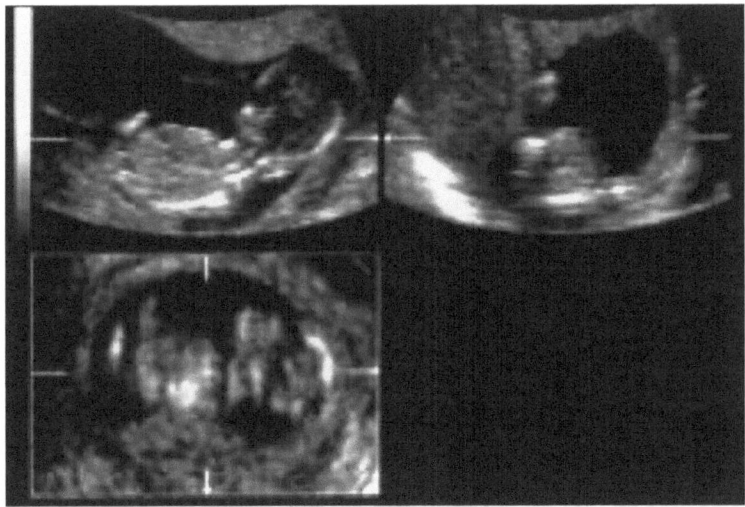

Fig. 2.23 NT measurements (2D) using Multi-planar reformatting (MPR) (Paul et al. 2001)

the described limitation in previous summaries section. In 2001 Paul et al. claims to measure NT thickness using 3D ultrasound imaging; however, their studies show that their measurement were conducted on re-slicing 2D ultrasound image from 3D multi-planar (MPR) visualization; sagittal, coronal and axial view plane. 2D measurements were taken rather than 3D thickness measurement, which in fact it makes no difference from conventional B-mode NT assessment. However, position of mid-sagittal plane selection can be known. Figure 2.23 illustrates their MPR measurement on 2D re-slice fetal images, rather than using 3D volume rendering measurement. Figure 2.24 illustrates the example of our 3D volume rendering measurement.

Furthermore, fully 3D acquisition systems are not widespread due to technological and economic reasons, especially in developing countries, and the majority of US scanners are freehand systems acquiring 2D B-scan images. Therefore, a software methodology for obtaining a 3D NT reconstruction and interactive visualization based on these systems is highly desirable.

2.7.1 Summary

With the rapid build-up of medical informatics technology and development, the demand on various sophisticated medical equipment are increasing dramatically. Nevertheless, many developing countries such as Malaysia are heavily depending on imported medical equipment, needless to say, the high cost reduces the treatment opportunity for the majority patients. This difficulty remains unsolved and limits patient category with only high income earners

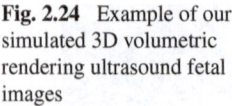

Fig. 2.24 Example of our simulated 3D volumetric rendering ultrasound fetal images

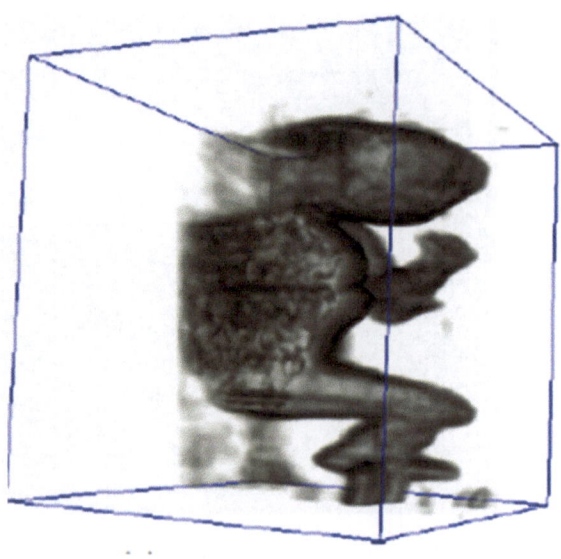

having easy access to benefits from high end medical technologies. This book proposed the techniques integrated with conventional 2D ultrasound systems in order to yield 3D interactive system dedicated for NT visualization system with no extra cost.

Generally, the ultrasound machines prices start at about RM 30,000 for 2D system and may range up to RM 700,000 for 3D or 4D system. A small heath center can purchase a simple B mode ultrasound machine on the very low end of this price range while a large hospital might pay nearly a million ringgit for a color machine with 4D images capability. The price difference between systems could rise up to 90 % high which depends largely on the level of technology. On the other hand, currently, many large-scaled hospitals, DICOM images are embedded in 3D reconstruction software which is similar to a class of large-scale image processing workstation or treatment planning system. A major drawback of the workstation system is the great consumption of computing processing. Therefore, to perform the task, demanding hardware configuration is needed and thus led to a rise of cost. Consequently, the establishment of such workstation is not affordable for most of the small-scaled hospitals. By virtue of its ineffectiveness, developer companies hardly to be benefited from the routine maintenance.

Besides, it is clear that 3D ultrasound has not been yet gained widespread of clinical acceptance. Most of the literatures presented reveal an emphasis on research projects, which extremely rarely and limitedly covered in routine protocol. It is used as a specialist laboratory tools for specific syndrome investigation, this is because the effort required to obtain high quality 3D ultrasound data often outweighs the potential benefits.

Another difficulty of 3D US fetal imaging is due to the fetus movement during scanning process. The automatic 3D probe mechanical movement may cover the images of fetus movement, especially in longer scanning time. This is the reason why 3D US system are not quantitatively reported for fetal biometrics measurement using existing commercial 3D probes except volume masking and its estimation. When medical personnel are not satisfied with the 3D scanning results, rescan is often performed. Hence, common 3D ultrasound probe operated at high frequencies for higher resolution will lower the penetration ability of the sound waves, i.e. 12 for 4D mode, 8 MHz for 3D mode. Apparently, proximal and distal NT layer beneath fetus neck is hardly assessable using high frequencies ultrasound setting. This limitation will be encountered in our research where conventional 3.5 MHz trans-abdominal probe is applied.

In present research, we have proposed a 3D ultrasound reconstruction and visualization for a specific ultrasound marker, namely nuchal translucency or NT, which is the important measurement parameter to assess the risk of trisomy 21 in early pregnancy. Measurements of NT were conducted in 3D spaces through the proposed state-of-art computerized algorithms. The proposed method has encountered the difficulty of current manual assessment method on using conventional B mode ultrasonic images. New visualization techniques have made it possible to create 2D cross sections that were not obtainable at regular scanning by processing a block of volume data (arbitrary slicing), presenting the surface of an object (surface shading), and looking into an object (transparency mode). Hence, it should be understood that due to the characteristics of NT marker anatomy, 3D thickness measurement on 3D structure with existing commercial ultrasound machine is not available. This project aims to design a non-invasive Trisomy 21 screening method easier for operator, at the same time, more robust removing the issues arise by FMF gold standard protocol. 3D visualization and measurement of ultrasound marker NT will be developed with the aim to improve the findings accuracy and consistency.

References

Aarnink, R. G., Huynen, A. L., Giesen, R. J. B., Delarosette, J., Debruyne, F. M. J., & Wijkstra, H. (1995). Automated prostate volume determination with ultrasonographic imaging. *Journal of Urology, 153*(5), 1549–1554.

Abele, H., Hoopmann, M., Wright, D., Hoffmann-Poell, B., Huettelmaier, M., Pintoffl, K., et al. (2010). Intra- and interoperator reliability of manual and semi-automated measurement of fetal nuchal translucency by sonographers with different levels of experience. *Ultrasound in Obstetrics and Gynecology, 36*(4), 417–422.

Abuhamad, A. (2005). Technical aspects of nuchal translucency measurement. *Seminars in Perinatology, 29*(6), 376–379.

Allott, C. P., Barry, C. D., Pickford, R., & Waterton, J. C. (1999). Volumetric assessment of carotid artery bifurcation using freehand-acquired, compound 3D ultrasound. *British Journal of Radiology, 72*(855), 289–292.

Arbeille, P., Eder, V., Casset, D., Quillet, L., Hudelo, C., & Herault, S. (2000). Real-time 3-D ultrasound acquisition and display for cardiac volume and ejection fraction evaluation. *Ultrasound in Medicine and Biology, 26*(2), 201–208.

Baba, K., Okai, T., Kozuma, S., & Taketani, Y. (1999). Fetal abnormalities: evaluation with real-time-processible three-dimensional US—preliminary report. *Radiology, 211*(2), 441–446.

Basset, O., Gimenez, G., Mestas, J. L., Cathignol, D., & Devonec, M. (1991). Volume measurement by ultrasonic transverse or sagittal cross-sectional scanning. *Ultrasound in Medicine and Biology, 17*(3), 291–296.

Bekker, M. N., Twisk, J. W. R., & van Vugt, J. M. G. (2004). Reproducibility of the fetal nasal bone length measurement. *Journal of Ultrasound in Medicine, 23*(12), 1613–1618.

Bernardino, F., Cardoso, R., Montenegro, N., Bernardes, J., & de Sa, J. M. (1998). Semiautomated ultrasonographic measurement of fetal nuchal translucency using a computer software tool. *Ultrasound in Medicine and Biology, 24*(1), 51–54.

Blaas, H. G., Eik-Nes, S. H., Berg, S., & Torp, H. (1998). In vivo three-dimensional ultrasound reconstructions of embryos and early fetuses. *Lancet, 352*(9135), 1182–1186.

Bonillamusoles, F., Raga, F., Osborne, N. G., & Blanes, J. (1995). Use Of 3-dimensional ultrasonography for the study of normal and pathological morphology of the human embryo and fetus—preliminary-report. *Journal of Ultrasound in Medicine, 14*(10), 757–765.

Braithwaite, J. M., & Economides, D. L. (1995). The measurement of nuchal translucency with transabdominal and transvaginal sonography—success rates, repeatability and levels of agreement. *British Journal of Radiology, 68*(811), 720–723.

Bulletins, A. C. o. P. (2007). Acog practice bulletin no. 77: Screening for fetal chromosomal abnormalities. *Obstetrics and gynecology, 109*(1), 217–227.

Catanzariti, E., Fusco, G., Isgrò, F., Masecchia, S., Prevete, R., & Santoro, M. (2009). A semi-automated method for the measurement of the fetal nuchal translucency in ultrasound images.

Cicero, S., Bindra, R., Rembouskos, G., Spencer, K., & Nicolaides, K. H. (2003a). Integrated ultrasound and biochemical screening for trisomy 21 using fetal nuchal translucency, absent fetal nasal bone, free Beta-Hcg and Papp-A At 11 To 14 weeks. *Prenatal Diag, 23*(4), 306–310.

Cicero, S., Dezerega, V., Andrade, E., Scheier, M., & Nicolaides, K. H. (2003b). Learning curve for sonographic examination of the fetal nasal bone at 11–14 weeks. *Ultrasound in Obstetrics and Gynecology, 22*(2), 135–137.

Cullen, M. T., Green, J. J., Reece, E. A., & Hobbins, J. C. (1989). A comparison of trans-vaginal and abdominal ultrasound in visualizing the 1st trimester conceptus. *Journal of Ultrasound in Medicine, 8*(10), 565–569.

Down, J. L. (1995). Observations on an Ethnic classification of idiots. *Mental Retardation, 33*(1), 54–56.

Downey, D. B., Nicolle, D. A., Levin, M. F., & Fenster, A. (1996). Three-dimensional ultrasound imaging of the eye. *Eye, 10*, 75–81.

Fenster, A., & Downey, D. B. (1996). 3-D ultrasound imaging: a review. *IEEE Engineering in Medicine and Biology Magazine, 15*(6), 41–51.

Fenster, A., Tong, S., Sherebrin, S., Downey, D. B., & Rankin, R. N. (1995). Three-dimensional ultrasound imaging. *Proceedings of SPIE, 2432*, 176–184.

Fenster, A., Downey, D. B., & Cardinal, H. N. (2001). Three-dimensional ultrasound imaging. *Physics in Medicine & Biology, 46*(5), R67–R99.

Gopal, A. S., Schnellbaecher, M. J., Shen, Z. Q., Akinboboye, O. O., Sapin, P. M., & King, D. L. (1997). Freehand three-dimensional echocardiography for measurement of left ventricular mass: in vivo anatomic validation using explanted human hearts. *Journal of the American College of Cardiology, 30*(3), 802–810.

Hafner, E., Philipp, T., Schuchter, K., Dillinger-Paller, B., Philipp, K., & Bauer, P. (1998). Second-trimester measurements of placental volume by three-dimensional ultrasound to predict small-for-gestational-age infants. *Ultrasound in Obstetrics and Gynecology, 12*(2), 97–102.

Howe, D., Wheeler, T., & Perring, S. (1994). Measurement of placental volume with real-time ultrasound in mid-pregnancy. *Journal of Clinical Ultrasound, 22*(2), 77–83.

Huether, C. A., Ivanovich, J., Goodwin, B. S., Krivchenia, E. L., Hertzberg, V. S., Edmonds, L. D., et al. (1998). Maternal age specific risk rate estimates for down syndrome among live births in whites and other races from Ohio and Metropolitan Atlanta, 1970–1989. *Journal of Medical Genetics, 35*(6), 482–490.

Hyett, J. A., Moscoso, G., & Nicolaides, K. H. (1995). Cardiac defects in 1st-trimester fetuses with trisomy 18. *Fetal Diagnosis and Therapy, 10*(6), 381–386.

Kagan, K. O., Wright, D., Spencer, K., Molina, F. S., & Nicolaides, K. H. (2008). First-trimester screening for trisomy 21 by free beta-human chorionic gonadotropin and pregnancy-associated plasma protein-A: Impact of maternal and pregnancy characteristics. *Ultrasound in Obstetrics and Gynecology, 31*(5), 493–502.

Kagan, K. O., Cicero S., et al (2009). Fetal nasal bone in screening for trisomies 21, 18 and 13 and Turner syndrome at 11–13 weeks of gestation. *Ultrasound in Obstetrics and Gynecology, 33*(3):259–264.

Kampmann, W., Walka, M. M., Vogel, M., & Obladen, M. (1998). 3-D sonographic volume measurement of the cerebral ventricular system: In vitro validation. *Ultrasound in Medicine and Biology, 24*(8), 1169–1174.

Kanellopoulos, V., Katsetos, C., & Economides, D. L. (2003). Examination of fetal nasal bone and repeatability of measurement in early pregnancy. *Ultrasound in Obstetrics and Gynecology, 22*(2), 131–134.

Kelly, I. M. G., Gardener, J. E., & Lees, W. R. (1992). 3-Dimensional Fetal Ultrasound. *Lancet, 339*(8800), 1062–1064.

Kornman, L. H., Morssink, L. P., Beekhuis, J. R., deWolf, B., Heringa, M. P., & Mantingh, A. (1996). Nuchal translucency cannot be used as a screening test for chromosomal abnormalities in the first trimester of pregnancy in a routine ultrasound practice. *Prenatal Diag, 16*(9), 797–805.

Laudy, J. A. M., Janssen, M. M. M., Struyk, P. C., Stijnen, T., Wallenburg, H. C. S., & Wladimiroff, J. W. (1998). Fetal liver volume measurement by three-dimensional ultrasonography: a preliminary study. *Ultrasound in Obstetrics and Gynecology, 12*(2), 93–96.

Lee, Y.-B., & Kim, M.-H. (2006). Automated ultrasonic measurement of fetal nuchal translucency using dynamic programming. In J. Martínez-Trinidad., J. A. Carrasco Ochoa., & J. Kittler. (Eds.), *Progress in Pattern Recognition, Image Analysis and Applications* (Vol. 4225, pp. 157–167), Berlin/Heidelberg: Springer.

Lee, A., Deutinger, J., & Bernaschek, G. (1995). 3-dimensional ultrasound—abnormalities of the fetal face in surface and volume rendering mode. *British Journal of Obstetrics and Gynaecology, 102*(4), 302–306.

Lee, Y.-B., Kim, M.-J., & Kim, M.-H. (2007). Robust border enhancement and detection for measurement of fetal nuchal translucency in ultrasound images. *Medical & Biological Engineering & Computing, 45*(11), 1143–1152.

Leshin, L. (1997). *Trisomy 21: The story of down syndrome*. Health Issues: D-S health Down Syndrome.

Magni, G., Cao, Q. L., Sugeng, L., Delabays, A., Marx, G., Ludomirski, A., et al. (1996). Volume-rendered, three-dimensional echocardiographic determination of the size, shape, and position of atrial septal defects: validation in an in vitro model. *American Heart Journal, 132*(2), 376–381.

Malone, F. D,. Ball, R. H., Nyberg, D. A., Comstock, C. H., Saade, G., Berkowitz, R. L., Dugoff, L., Craigo, S. D., Carr, S. R., Wolfe, H. M., Tripp, T., D'Alton, M. E., & Consortium, F. R. (2004). First-trimester nasal bone evaluation for aneuploidy in the general population. Obstetrics and gynecology, *104*(6), 1222–1228.

Martin, R. W., Bashein, G., Detmer, P. R., & Moritz, W. E. (1990). Ventricular volume measurement from a multiplanar transesophageal ultrasonic-imaging system—an invitro study. *IEEE Transactions on Biomedical Engineering, 37*(5), 442–449.

Moratalla, J., Pintoffl, K., Minekawa, R., Lachmann, R., Wright, D., & Nicolaides, K. H. (2010). Semi-automated system for measurement of nuchal translucency thickness. *Ultrasound in Obstetrics and Gynecology, 36*(4), 412–416.

Nagdyman, N., Walka, M. M., Kampmann, W., Stöver, B., & Obladen, M. (1999). 3-D ultrasound quantification of neonatal cerebral ventricles in different head positions. *Ultrasound in Medicine and Biology, 25*(6), 895–900.

Nathan, M. S., Seenivasagam, K., Mei, Q., Wickham, J. E. A., & Miller, R. A. (1996). Transrectal ultrasonography: Why are estimates of prostate volume and dimension so inaccurate? *British Journal of Urology, 77*(3), 401–407.

Nicolaides, K. H., Azar, G., Byrne, D., Mansur, C., & Marks, K. (1992). Fetal nuchal translucency: ultrasound screening for chromosomal defects in first trimester of pregnancy. *BMJ*,
304(6831), 867–869.

Nicolaides, K. H., Brizot, M. L., & Snijders, R. J. M. (1994). Fetal nuchal translucency—ultrasound screening for fetal trisomy in the first trimester of pregnancy. *British Journal of
Obstetrics and Gynaecology, 101*(9), 782–786.

Nicolaides, K., Sebire, N., & Snijders, R. (1999). *The 11–14 weeks scan: the diagnosis of fetal
abnormalities*. New York, NY: Parthenon Publishing.

Nicolaides, K. H., Kagan, K. O., Wright, D., Baker, A., & Sahota, D. (2008). Screening for trisomy 21 by maternal age, fetal nuchal translucency thickness, free beta-human chorionic
gonadotropin and pregnancy-associated plasma protein-A. *Ultrasound in Obstetrics and
Gynecology, 31*(6), 618–624.

Ofili, E. O., & Nanda, N. C. (1994). 3-dimensional and 4-dimensional echocardiography.
Ultrasound in Medicine and Biology, 20(8), 669–675.

Pandya, P. P., Brizot, M. L., Kuhn, P., Snijders, R. J. M., & Nicolaides, K. H. (1994). First-
trimester fetal nuchal translucency thickness and risk for trisomies. *Obstetrics and
Gynecology, 84*(3), 420–423.

Pandya, P. P., Altman, D. G., Brizot, M. L., Pettersen, H., & Nicolaides, K. H. (1995).
Repeatability of measurement of fetal nuchal translucency thickness. *Ultrasound in
Obstetrics and Gynecology, 5*(5), 334–337.

Paul, C., Krampl, E., Skentou, C., Jurkovic, D., & Nicolaides, K. H. (2001). Measurement
of fetal nuchal translucency thickness by three-dimensional ultrasound. *Ultrasound in
Obstetrics and Gynecology, 18*(5), 481–484.

Perona, P., & Malik, J. (1990). Scale-space and edge detection using anisotropic diffusion.
Pattern analysis and machine intelligence. *IEEE Transactions On, 12*(7), 629–639.

Pohls, U. G., & Rempen, A. (1998). Fetal lung volumetry by three-dimensional ultrasound.
Ultrasound in Obstetrics and Gynecology, 11(1), 6–12.

Rankin, R. N., Fenster, A., Downey, D. B., Munk, P. L., Levin, M. F., & Vellet, A. D. (1993).
Three-dimensional sonographic reconstruction: techniques and diagnostic applications.
American Journal of Roentgenology, 161(4), 695–702.

Roberts, L. J., Bewley, S., Mackinson, A. M., & Rodeck, C. H. (1995). First trimester fetal
nuchal translucency—problems with screening the general-population. 1. *British Journal* of
Obstetrics and *Gynaecology 102*(5), 381–385.

Rosenfield, K., Boffetti, P., Kaufman, J., Weinstein, R., Razvi, S., & Isner, J. M. (1992). 3-dimen-
sional reconstruction of human carotid arteries from images obtained during noninvasive
B-mode ultrasound examination. *American Journal of Cardiology, 70*(3), 379–384.

Salustri, A., & Roelandt, J. (1995). Ultrasonic 3-dimensional reconstruction of the heart.
Ultrasound in Medicine and Biology, 21(3), 281–293.

Schild, R. L., Wallny, T., Fimmers, R., & Hansmann, M. (1999). Fetal lumbar spine volumetry
by three-dimensional ultrasound. *Ultrasound in Obstetrics and Gynecology, 13*(5), 335–339.

Snijders, R. J. M., Noble, P., Sebire, N., Souka, A., Nicolaides, K. H., & Grp, F. M. F. F. T.
S. (1998). Uk multicentre project on assessment of risk of trisomy 21 by maternal age
and fetal nuchal-translucency thickness at 10–14 weeks of gestation. *Lancet, 352*(9125),
343–346.

Snijders, R. J. M., Sundberg, K., Holzgreve, W., Henry, G., & Nicolaides, K. H. (1999). Maternal
age- and gestation-specific risk for trisomy 21. *Ultrasound in Obstetrics and Gynecology,
13*(3), 167–170.

Souka, A. P., Krampl, E., Bakalis, S., Heath, V., & Nicolaides, K. H. (2001). Outcome of pregnancy in chromosomally normal fetuses with increased nuchal translucency in the first trimester. *Ultrasound in Obstetrics and Gynecology, 18*(1), 9–17.

Wee L. K., Lim M., & Supriyanto, E. (2010a). Automated Risk Calculation For Trisomy 21
Based On Maternal Serum Markers Using Trivariate Lognormal Distribution. *Proceedings
of the 12th WSEAS International Conference on Automatic Control, Modelling & Simulation
(ACMOS 2010)*. pp. 327–332.

Taipale, P., Hiilesmaa, V., Salonen, R., & Ylostalo, P. (1997). Increased nuchal translucency as a marker for fetal chromosomal defects. *New England Journal of Medicine, 337*(23), 1654–1658.

Terris, M. K., & Stamey, T. A. (1991). Determination of prostate volume by transrectal ultrasound. *Journal of Urology, 145*(5), 984–987.

Ville (2010). Semi-automated measurement of nuchal translucency thickness: Blasphemy or oblation to quality? *Ultrasound in Obstetrics & Gynecology. 36*(4), 400–403.

Wald, N. J., George, L., Smith, D., Densem, J. W., & Pettersonm, K. (1996). On behalf of the International Prenatal Screening Research, G. Serum Screening For Down's Syndrome Between 8 And 14 Weeks Of Pregnancy. *BJOG: An International Journal of Obstetrics & Gynaecology, 103*(5), 407–412.

Wald, N. J., Rodeck, C., Hackshaw, A. K., Walters, J., Chitty, L., Mackinson, A. M., & Group, S. R. (2003). First and second trimester antenatal screening for down's syndrome: the results of the serum, urine and ultrasound screening study (suruss). *Health Technology Assessment (Winchester, England), 7*(11), 1–77.

Wee, L. K., Miin, L., & Supriyanto, E. (2010). Automated trisomy 21 assessment based on maternal serum markers using trivariate lognormal distribution. *WSEAS Transactions on Systems, 9*(8), 844–853.

Wee, L. K., Chai, H. Y., & Supriyanto, E. (2011). Computerized nuchal translucency three dimensional reconstruction, visualization and measurement for trisomy 21 prenatal early assessment. *International Journal of Physical Sciences, 6*(19), 4640–4648.

Zosmer, N., Souter, V. L., Chan, C. S. Y., Huggon, I. C., & Nicolaides, K. H. (1999). Early diagnosis of major cardiac defects in chromosomally normal fetuses with increased nuchal translucency. *BJOG: An International Journal of Obstetrics & Gynaecology, 106*(8), 829–833.

Chapter 3
Designs and Implementation of Three Dimensional Nuchal Translucency

Abstract General principles of 3D reconstruction are composed of two steps. Firstly, the captured successive 2D image slices are arranged precisely with real spatial positions, resulting in volumetric data. This will be followed by ray casted volume rendering techniques, hybrid with semi-auto 3D segmentation and interactive virtual slider for 3D NT measurements. The methodology entails the composite function to visualize the explicit internal marker structure. 3D image diffusion is utilized as preprocessing technique before segmentation, 3D seeded growing is utilized to segment the 3D NT structure followed by its interactive measurement. Details of numerical measurement analysis shall be discussed in Chap. 4. This chapter describes the experimental design and implementation which includes research materials, data sources acquisitions and manipulation, 3D reconstruction and scanning techniques, volume rendering visualization techniques, 3D image preprocessing diffusion, semi-automated 3D segmentation designs, virtual extraction, hybrid visualization and interactive measurements.

3.1 Introduction

In present research, multiple B-mode ultrasonic fetal images in successive order are collected from collaborator hospital with help from professional healthcare personnel. Specific protocols were prepared for sonographer and detail explanations are performed prior to clinical patient ultrasound scanning at the hospital. The aim of these clinical data collection is to build a proof-of-principle technique for 3D nuchal translucency thickness measurement in semi-automated function, in order to encounter all the critical limitations of existing conventional 2D measurement method. The entire work frame design is shown in Fig. 3.1. The process begins with clinical data acquisition, data manipulation for 3D reconstruction, 3D segmentation, 3D visualization and 3D measurement in high flexibility function. Interactivity of the developed algorithm in 3D simulation was built as callback command. It stands as better solution for our specific ultrasonic marker measurement as compared to existing commercial ultrasound imaging mode. Several

K. W. Lai and E. Supriyanto, *Detection of Fetal Abnormalities Based on Three Dimensional Nuchal Translucency*, SpringerBriefs in Applied Sciences and Technology, DOI: 10.1007/978-981-4021-96-8_3, © The Author(s) 2013

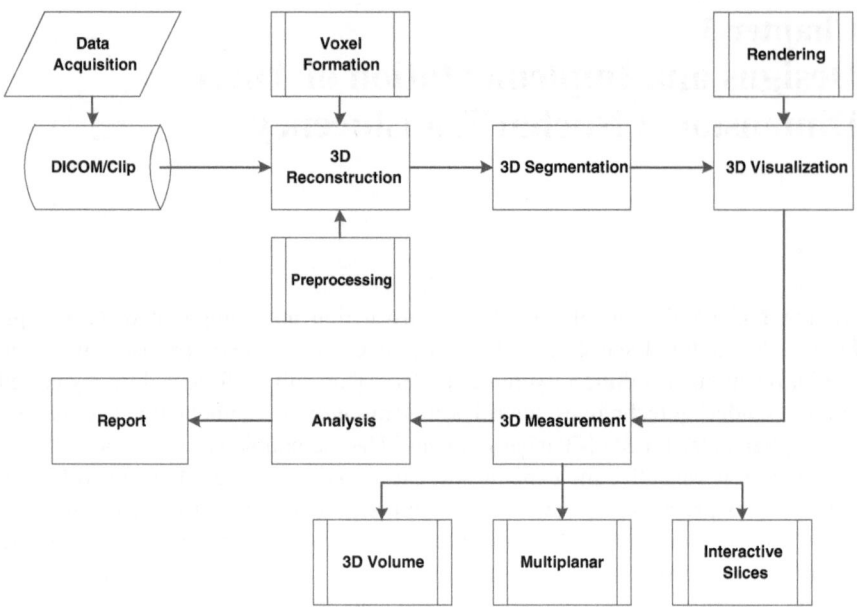

Fig. 3.1 Three dimensional nuchal translucency assessment design

developed unique visualization include hybrid visualization of MPR with volume rendering integrated with virtual sliders; this is not found from the existing commercial ultrasound imaging mode.

The proposed 3D NT marker assessment method have put their benefit forward includes: (a) the acquisition of the required volume scanning areas allows clinician to view desired cross section plane regardless the position of fetal, which is otherwise not possible due to the physical constraints of the prenatal scanning process. For example, 2D slices in parallel with the skin surfaces are not possible from conventional 2D ultrasound systems. (b) Optimal mid-sagittal plane may be overlooked; hand-eye coordination of clinician can be replaced with direct 3D scene visualization and plane selection. Also, in certain complicated cases, 2D images acquired by sonographer could arise confusion in diagnostics decision and therefore, expertise consultation by medical personnel is required. If a comprehensive 3D ultrasound data set is available then it can be viewed and the most diagnostically significant plane shall be navigated. This method could save time and costs of re-scanned procedure and avoid inconvenience of the case where experts do not satisfy the saved 2D slices, which is not possible to be changed or modified. (c) 3D reconstructed NT visualization and measurement could be more informative as compared to conventional 2D cross sectional B-mode images. The proposed findings are much more easier to be interpreted which in turn helps for operators who are less experiences, or non-FMF certificate holder for NT marker measurement. The above issues strongly suggest the need for an easy-to-use, practical and low-cost quantitative 3D ultrasound acquisition method for nuchal translucency. (d) The proposed technique may upgrade existing 2D ultrasound systems to 3D systems with minimal cost. This is much more beneficial in developing countries where 2D systems are widely implemented rather than 3D systems at extremely high cost. (e) It also appears as an alternative to other invasive systems (maternal serum markers and amniocentesis) for Trisomy 21 detection, as described in section Introduction.

3.2 Data Acquisition

The commercial 2D ultrasound systems used in this research is Siemens Acusons (Mountain View, CA, USA) at BMTI, Technische Universitat Ilmenau; Kontron (Sigma 330 Expert) at Universiti Teknologi Malaysia. The implemented ultrasound probe is convex abdominal transducer type with frequency 3.5 MHz. The tested ultra-sound fetal training phantom is CIRS model 065-20. The ultrasound machine used at collaborator hospital is GE Voluson with same type of transducer probe. Ethics consent application is needed prior to collect patient scanning data at the hospital. The application is submitted and approved by medicine ethics committee (Human) Hospital Universiti Sains Malaysia (HUSM). Figure 3.2 shows the application flow of ethics approval for clinical ultrasound data collection within the research periods.

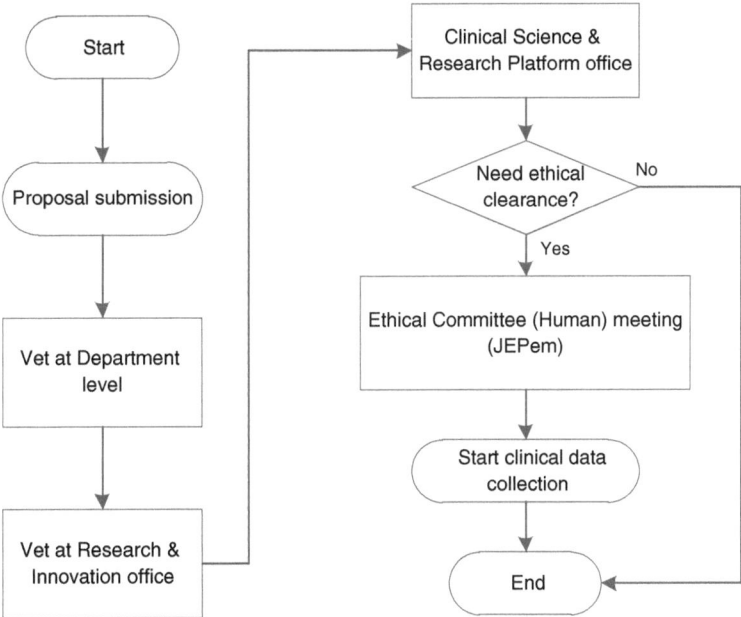

Fig. 3.2 Application flow ethics approval HUSM for clinical ultrasound data collection at hospital

The scanned subject in the present studies is pregnant women with first trimester pregnancies or between 11 until 13 weeks plus 6 days. Some collected data include early second trimester of pregnancies up to 20 weeks. The target research ultrasonic marker is nuchal translucency (NT) utilized by conventional 2D B-mode prenatal ultrasound scan protocol. The motivation of NT marker assessment is to classify the risk of fetus with Trisomy 21 during early pregnancy. Patient populations include Malaysian with only singleton gestation screening operated by two experienced senior sonographers served in public hospital. For current clinical assessment, three trials 2D ultrasonic NT thickness measurement for each patient were taken, and mean average is treated as final finding for diagnostics decision. These medical measurements will be treated as standard references for comparison and correlation with the proposed 3D findings. Figure 3.3 shows part of the data collection for the same patient with anonymous privacy data, which illustrates the existing 2D methodological problem of its current measurement protocol: inconsistency of 2D manual thickness measurement. For instance, these three trials measurement have maximum deviation up to 0.21 mm in this example.

The deviation measurement can be larger, as the operator changes the position of sagittal view during prenatal screening. Selecting the correct sagittal plane is crucial, which is very subjective and relied on personal experiences and technical skills. Nevertheless, 2D ultrasonic assessment is prone to human error, and subject to inter and intra-observer variability (Bekker et al. 2004; Abuhamad 2005; Chen et al. 2010).

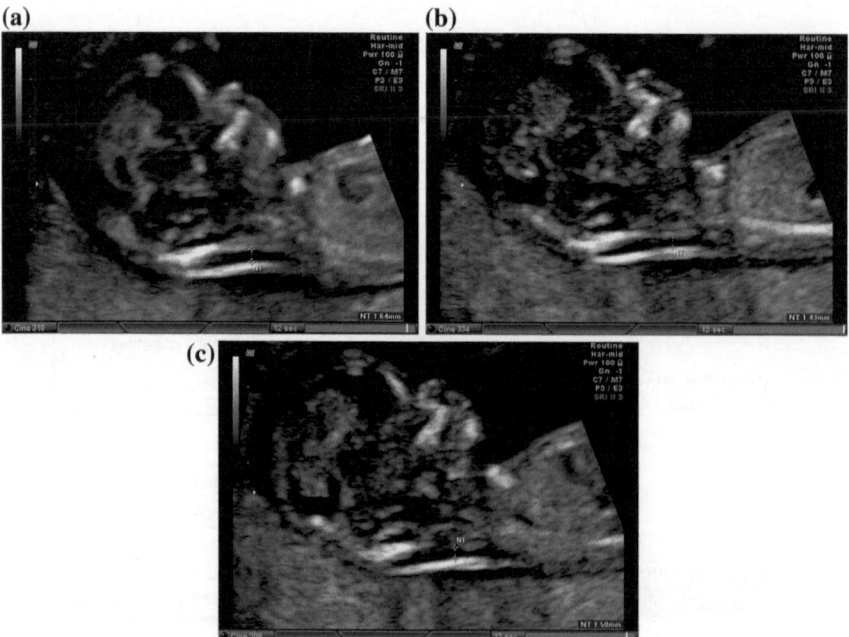

Fig. 3.3 Three trials manual B-mode Ultrasound sonogram measurement results large variability. **a** NT thickness = 1.64 mm, **b** NT thickness = 1.43 mm, **c** NT thickness = 1.50 mm

(a)   **(b)**

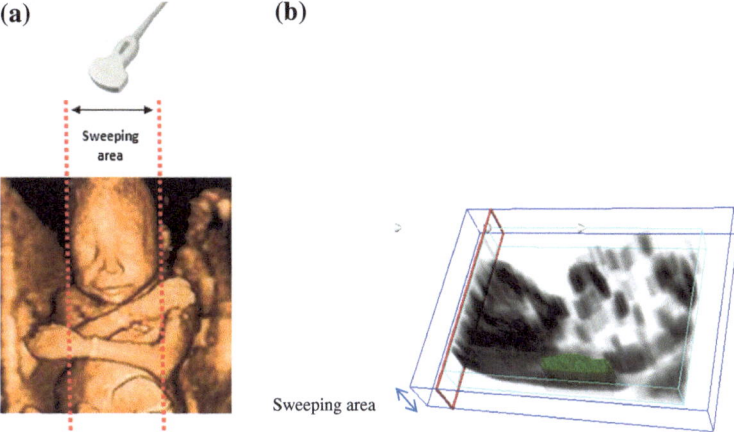

Fig. 3.4 Ultrasound probe scanning area in linear position movement. **a** Example of sweeping area on ultrasound fetal training phantom. **b** Experimental simulation of 3D reconstruction and volume rendering

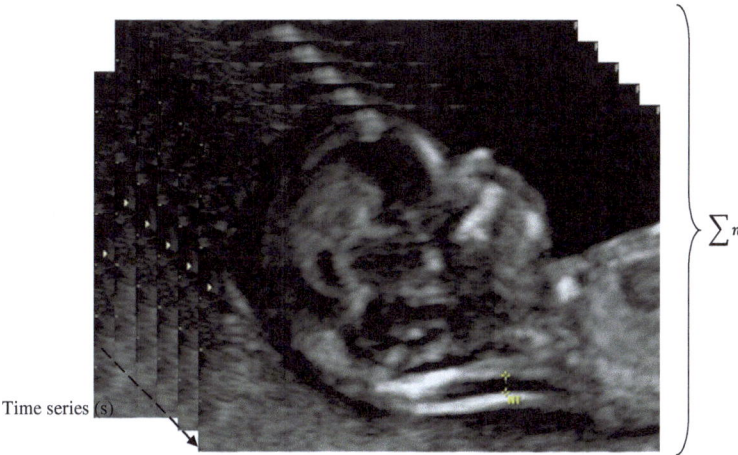

Fig. 3.5 Ideal 2D image acquisitions for multi-frames recording

Mid sagittal view of the fetal profile must be covered by moving the transducer probe from side to side so that the inner edges of the two thin echogenic lines that border the NT layer is obtained. Figure 3.4 illustrates the proposed moving transducer area as conducted by sonographers. The magnification of the image should be at least 75 % zooming such that the head and thorax region occupy full screen of the image in the neutral position, whenever it is possible. The ultrasound gain setting remained unchanged throughout the transducer movement. The ultrasound images were obtained as the sequence of moving pictures. The range of acquisition time is between 7 and 12 s. In other words, total numbers of recorded frames are between 175 and 300 consecutive images at frame rate 25/s. Figure 3.5 shows the ideal sequences of 2D ultrasound data collection at our collaborator hospital with anonymous patients' private information.

Total number of frames can be calculated using;

$$\Sigma n = f \times s \qquad (3.1)$$

where,

   n = total frames;
   f = frame rate;
   s = acquisition time (s).

3.3 DICOM and Clips Storage

Storage of ultrasound data can be two forms in present studies, DICOM and moving clips format. DICOM is also known as Digital Imaging and Communications in Medicine, which is specially used for medical digital imaging and communications, jointly developed by American College of Radiology (ACR) and National Electrical Manufacturers Association (NEMA). The collected multi-frames DICOM files and sequencing clips are stored in 8 bit, and digital unsigned characteristics with grey scale level between 0 and 255. During our early research stage, we have developed hardware assembly for analog–digital ultrasound image acquisition for synchronizing image recording. Figures 3.6 and 3.7 shows the block diagram of the developed hardware assembly and experimental setup respectively. Figure 3.8 shows the designs of the hardware casing.

During the acquisition process at hospital, there are total 2 keynotes need to be complied by sonographer or operator. First, it is clear that freehand ultrasound data acquisition might subject to unequaled frames position in transducer moving direction. External position tracking devices attaching on clinical usage ultrasound system in hospital is prohibited and it will not be in our case. Therefore, sonographers have to move the scanning transducer in linear position at a regular speed, as precise as possible. This assumption can be realized with experienced health care professional scanning at the good sagittal view recording without affecting axial resolution and lateral resolution after 3D reconstruction; this is because NT thickness is the depth parallel (axial direction) to ultrasound beam form when abdominal probes are positioned correctly and followed by FMF protocol. The nomenclature resolution equals the original 2D images, but perpendicular to the acquired 2D image planes, it equals to the elevational resolution of the transducer.

Second, the moving distance of scanning transducer is extremely small during first trimester NT assessment; less than few millimeters. In normal condition of first trimester gestation age (GA), the crown rump length (CRL) ranges between 45 and 84 mm; biparietal diameter (BPD) ranges between 16 and 24 mm. Therefore, the moving distance of sagittal probe position is less than 10 mm on the grounds

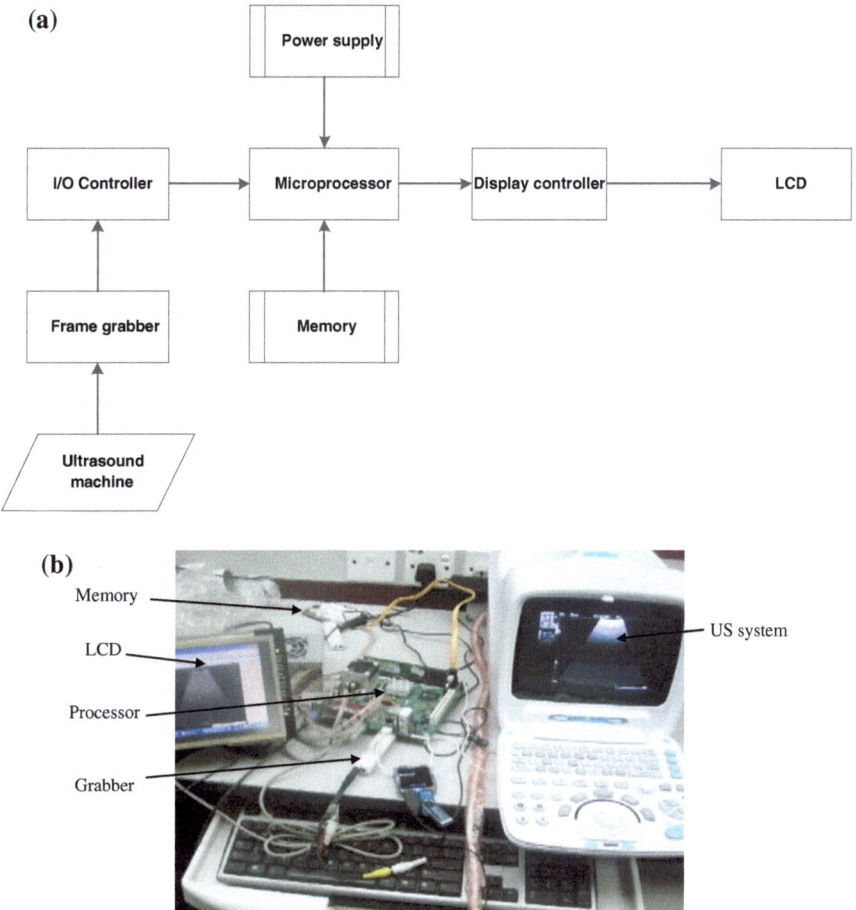

Fig. 3.6 Developed hardware for synchronize data acquisition. **a** Block diagram. **b** Assembly hardware with low end 2D ultrasound system

of small NT profile. This extreme short moving scanned distance aims to reduce or neglect the potential interference, for example, fetal movement, which is yet the complicated difficulty for existing 3D US imaging probes. Figure 3.9 illustrates the example measurement of BPD in 2D US system and present research method respectively.

Hence, trans-vaginal probe is not implemented in current investigation as it does not cover in routine scanning protocol. Generally, it is applied for exceptional case such as patients faced with obesity problem, and difficult fetal position. This also explains why existing commercial 3D ultrasound system is not applied on depth structure NT profile assessment. Common 3D ultrasound probe operated at high frequencies for higher resolution will lower the penetration ability of the sound waves, i.e. 12 MHz for 4D mode, 8 MHz for 3D

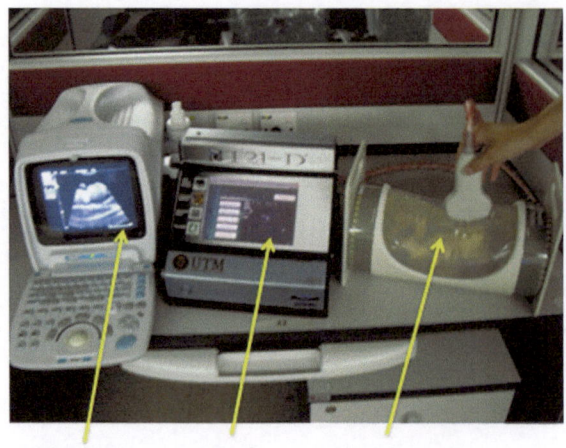

US machine   T21-D   Fetal phantom

Fig. 3.7 Experimental setup of 2D image frame acquisition using ultrasound fetal training phantom (*right*); acquisition hardware (*middle*); 2D low end ultrasound system (*left*)

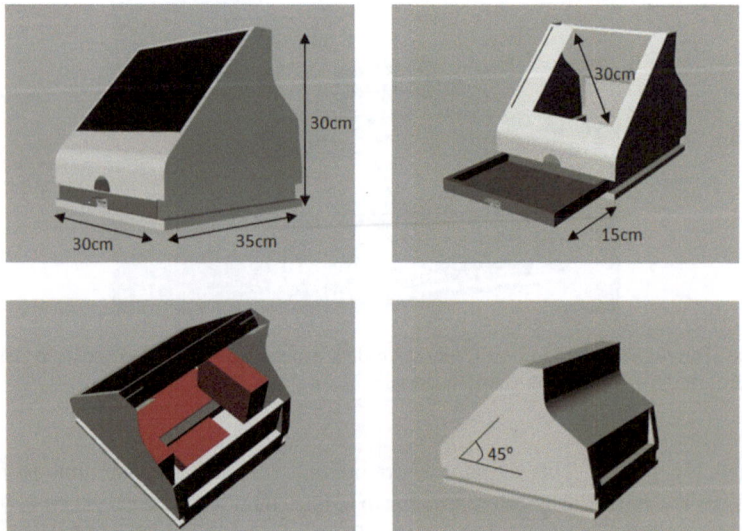

Fig. 3.8 AutoCAD design drawing for the hardware assembly

mode. Apparently, proximal and distal NT layer beneath fetus neck is hardly assessable using high frequencies ultrasound setting. This limitation will be encountered in our research where conventional 3.5 MHz trans-abdominal probe is applied. Figure 3.10 shows different common available type of ultrasound probe.

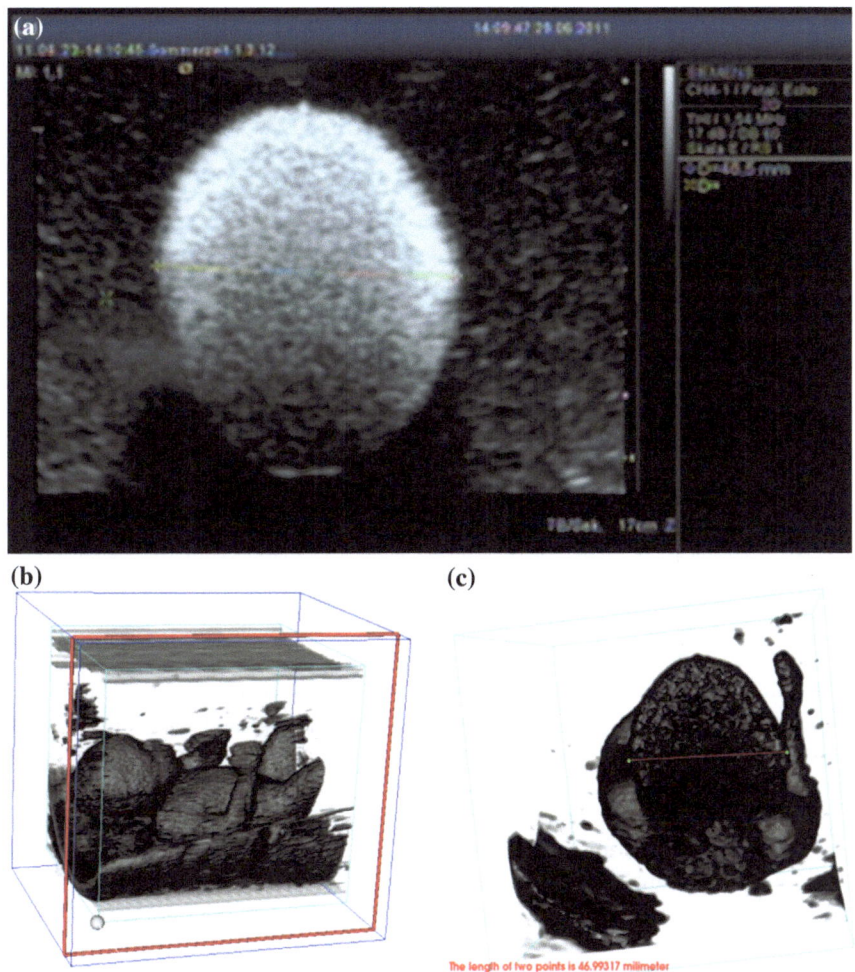

Fig. 3.9 Example BPD measurements (**a**) 2D marker measurement using 2D commercial US system (**b**), (**c**) 3D marker measurement using present research method

3.4 3D Nuchal Translucency Reconstruction and Visualization

In this research, the proposed 3D reconstructions and computerized rendering program are implemented on the basis of some external open-sources object-oriented libraries. The integrated IDE environment is Microsoft Visual Studio 8.0 in Window 7 using C++ language programming, where the initial project file can be built using CMake 2.80 while configuring external libraries including OpenCV 2.2, VTK 5.4, OpenIGTLink 1.0 and ITK 3.2. These external libraries act as parallel resources for

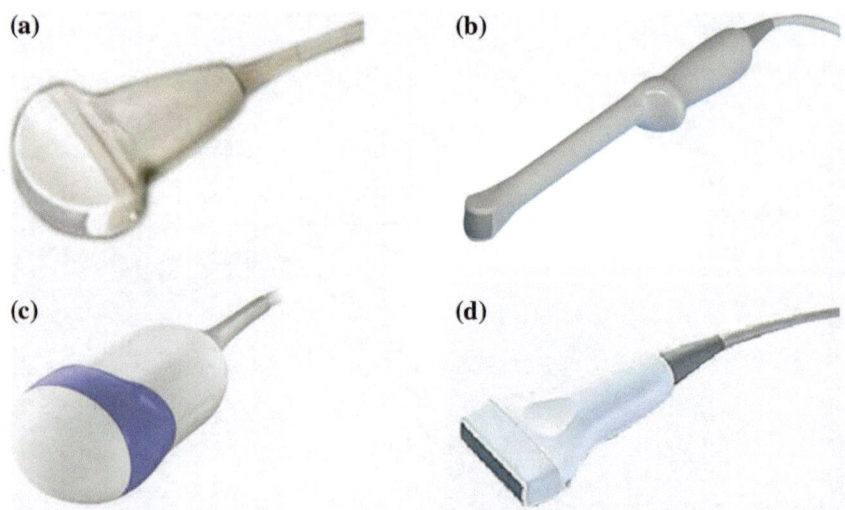

Fig. 3.10 Ultrasound probe types. **a** Trans-abdominal convex transducer. **b** Trans-vaginal transducer. **c** 3D transducer. **d** Linear array transducer

the developed support classes. They have excellent structure and operating mechanism, causing it to be widely used in the international visualization research area. It is strictly designed according to object-oriented concept and taken into account the simplicity of compiled programming and interpreter programming. To encounter the compatibility and ease of maintenance and upgrades, it utilizes the standard programming language environment, such as C++, TCL, Java or Python code to encapsulate common visualization algorithms. It can be operated in Windows, UNIX, Linux or other operating systems independently.

In our investigation, the developed algorithm can achieve a variety of interactive visualization and measurement simulation by developing new part of library class, and adjusting the appropriate means of implementation and execution of code embedded in personalized coding. Since the implementation of developed class library contains high flexibility and efficiency in real time interactive simulation, we decided to use C++ programming language for the proposed 3D ultrasound marker investigation.

In general, there are two main principles of 3D ultrasound reconstruction and visualization, which are surface and volume rendering (Fig. 3.11). Our previous published work (Wee et al. 2011) described the details of data extraction for fetal ultrasound contour surface rendering by using simplified marching cubes method. However, our early research findings show that the important structure and topology of NT will not be seen if this rendering technique is applied on reconstructed 3D fetal images. The brief explanation of surface rendering is as below (Fig. 3.12); the algorithm idea is to construct 3D data field and find out isosurface voxels. Position of triangle vertices with equivalent large number of triangles face will be determined through linear interpolation method. In

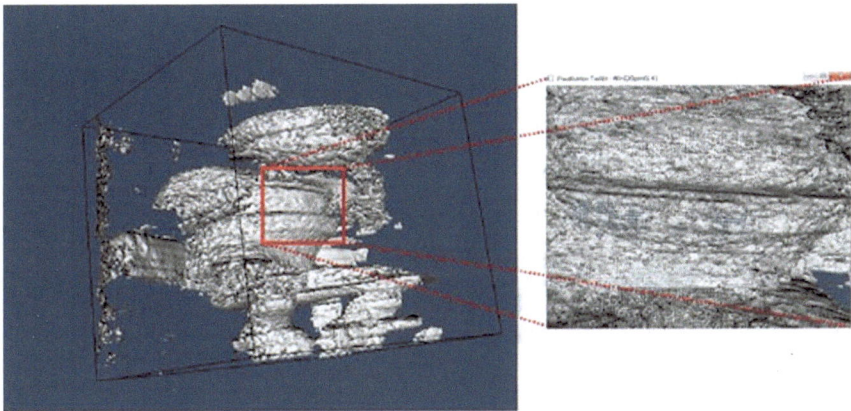

Fig. 3.11 Simulation result of 3D surface rendering, closed view (RHS) shows surface composed by triangles strips

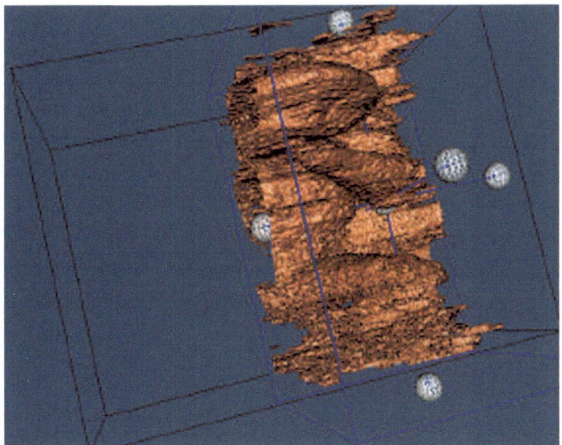

Fig. 3.12 Simulation result surface rendering

functional processes of marching cube (MC) 3D reconstruction, algorithm will read DICOM image file format in series, and then it will cut pictures into a collection of small cubes from volume element. MC algorithm will access each cube to generate equivalent collection of triangles. Unfortunately, standard MC algorithm used to generate an excessive number of triangles, causes follow-up treatment become a bottleneck, so simplification of triangular grid appears essentially vital. Therefore, we have proposed vertex merging algorithm to reduce generated triangle number. It aims to replace representative of verticals, which can be combined to form a new vertex in next round of consolidation. When two vertices satisfy the conditions, edges of two vertices will shrink to a point, where triangle vertices are deleted. If conditions are not met, triangle

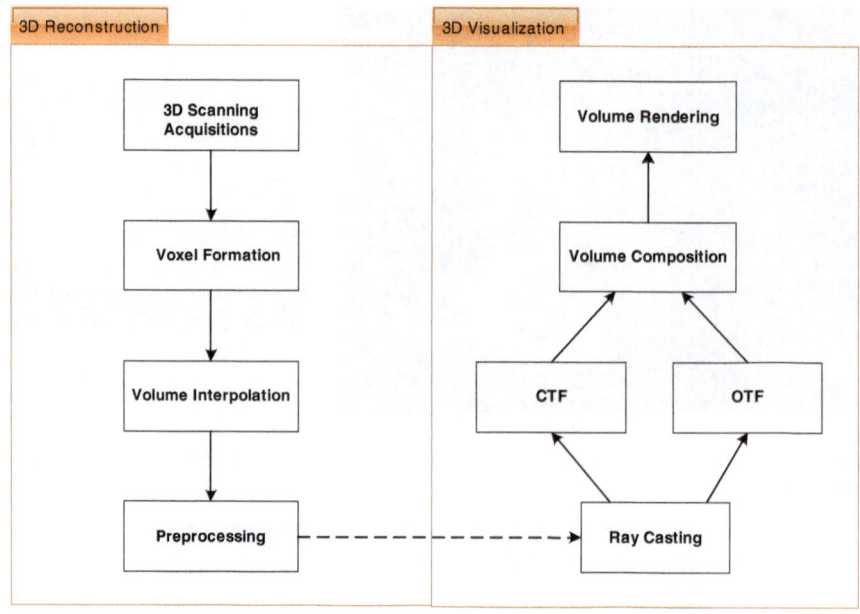

Fig. 3.13 Main pipeline mechanism design for 3D NT reconstruction and visualization

is retained and it will continue to determine next vertex of a triangle situation. Assuming that two vertices of triangle V_i, V_j angle between vectors $\alpha(V_i, V_j)$, a minimum angle constraint is set as;

$$\alpha(V_i, V_j) = \left\lceil \frac{N_i \cdot N_j}{|N_i| |N_j|} \right\rceil \tag{3.2}$$

where N_i, N_j denote normal vector of vertex V_i, V_j.

This approach will represents anatomical structures by simple surface boundary, where important image information including subtle NT anatomical features and its texture is lost; therefore, we have designed the 3D volumetric reconstruction pipeline mechanism dedicated for NT features visualization and analyzing system, as shown in Fig. 3.13. The main pipeline design mechanism for NT volume rendering is relatively similar to surface rendering, but it is using the elementary voxel unit of volume data and data manipulated by composite ray casting techniques.

3D reconstruction and visualization are separated into two distinct processes in our mechanism pipeline design. For 3D reconstruction, the process starts from scanning acquisition in order to form 3D voxel from the B-mode image pixels. The voxel based presentations are formed by linear interpolation from 2D pixel to 3D voxel. Each individual voxel values are determined by their respective pixel grey level intensity. 3D diffusion is manipulated as the preprocessing technique in order to enhance the quality of volumetric rendering. For 3D visualization, the casted voxel rays are integrated with opacity transfer function (OTF) and color transfer function (CTF) to form the composite rendering. The 3D visualization processes end with volume rendering channel.

The difference of volume rendering in present studies is to access the object's internal information, in our case, we will simulate the 3D nuchal translucency analyses in prominent way rather than two weak echogenic lines on 2D B-mode ultrasound images. It is followed by the NT thickness measurement using Euclidean distance equation in 3D formation. Figure 3.14 illustrates the overall reconstruction and visualization processing for 3D NT images. The proposed approach preserves all the information originally presents in the acquired 2D images, thus, by taking suitable cross-section plane or volume from the 3D data, as shown in the figure, the original 2D images can be recovered while new view plane not found from the original data set can be generated furthermore.

Several rendering techniques are proposed for NT marker investigation at a specific optimum plane since all the original image information is preserved. Operator can choose the suitable technique for displays the NT features to the best advantage. Operator may continue apply the 3D SRG segmentation and classify the NT fold thickness in volume-based rendering operations, as shown in Fig. 3.15.

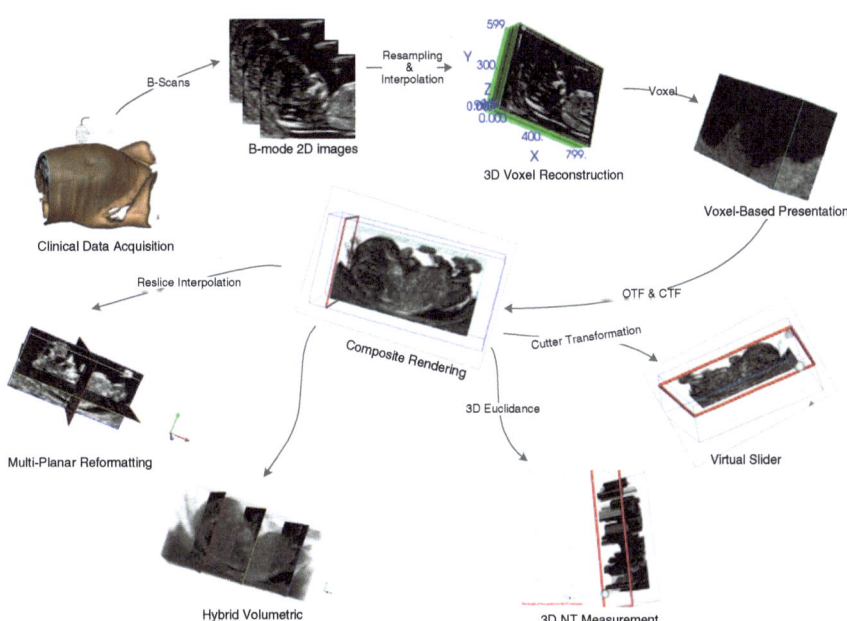

Fig. 3.14 Overall 3D reconstruction and visualization process

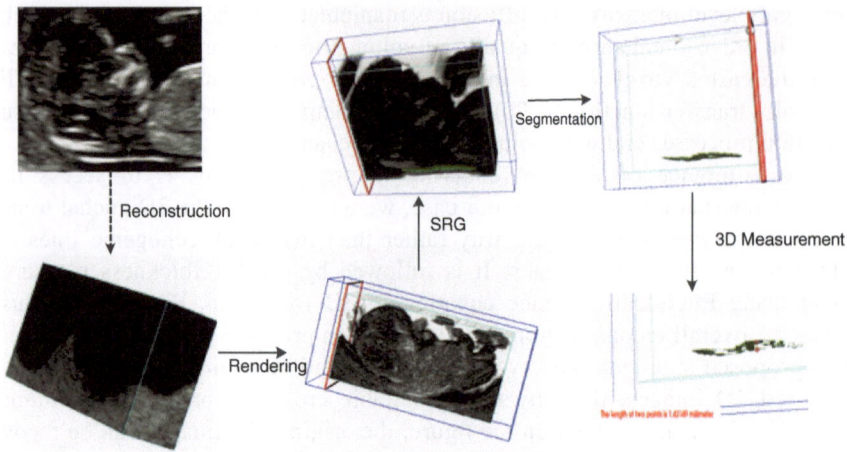

Fig. 3.15 Volumetric 3D nuchal translucency SRG based segmentation

3.4.1 Three Dimensional Scanning Acquisitions

The typical commercial 3D ultrasound system implementing 1D ultrasound transducer to acquire a series of 2D cross-section images, and forming 3D images through several restoration manipulations. Based on the literatures (Nelson and Pretorius 1998; Gee et al. 2003; Fenster and Downey 2000; Gee et al. 2004), scanning approaches can be categorized into three different approaches: mechanical scanners, free hand scanning with or without position trackers, and 2D transducer arrays, as shown in Fig. 3.16.

The existing developed mechanical scanner approaches consists of three basic sweeping protocols, includes rotation, tilt and linear probe movements controlled by stepper motor. Rotation simply means the probe motion rotates about an axis in axial direction at least of an angle 180°, tilt indicates the series of 2D images are collected at regular angular intervals as the transducer is tilted about an axis parallel to the transducer face, and linear probe movement collecting successive 2D images

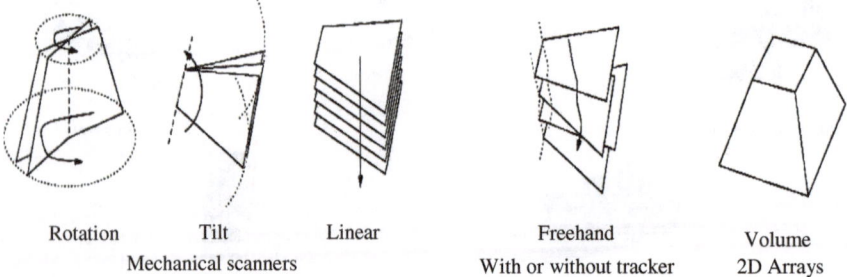

Fig. 3.16 Various 3D ultrasound scanning approaches

in parallel route at regular spatial intervals. The geometric parameters include relative positions and orientations of recorded 2D images are known accurately with the predefined stepper motor controller. The advantages of this approaches is the feasibility for operator or healthcare professional rather than engineer. However, the major drawbacks are the bulky 3D probes, which are larger and much heavier than conventional probes. It also requires purchase of special 3D ultrasound system to support and interface with this mechanical scanner. Besides, it is also impossible to scan any larger interest volume than the mechanics housing designs itself. This is due to the coordinate reference frame is based on the static stepper motor housing.

The previous section described the scanning approaches implementing conventional 1D array transducer; it requires the mechanical or manual movement control by sweeping across the scan object. Though, 2D arrays transducer offers spontaneous real time 3D ultrasound visualization. This approach requires the ultrasonic probe to remains static and it accesses the 3D volume data of subject interest directly. Nevertheless, the practical engineering of 2D planar array involves large number of array elements, each array element requires configuration to appropriate channels. Therefore, various issues appear as limitation such as technical complexity, expensive element in small size, numerous electronic connections, low yield of manufacturing process, and high system cost (Fenster and Downey 1996). These factors have slowed down the development and acceptance of this technology.

Freehand scanning approaches consist of two separate protocols, either with or without attaching a position tracker on the ultrasound transducer. For freehand scanning with trackers, it may include four major sensing instruments using acoustics sensors (King et al. 1994; Altmann et al. 1997), optical sensors (Trobaugh et al. 1994), electromagnetic sensors (Meairs et al. 2000; Berg et al. 1999; Gobbi et al. 1999, 2000; Gobbi and Peters 2002) and mechanical arms (Ohbuchi and Fuchs 1991) respectively. The ultrasound probe is manipulated in arbitrary manner while the sensors tracked the position and orientation of each successive 2D image. This approach provides high flexibility with no mechanical constrains as faced by mechanical scanner approaches, for example the bulky hardware or motor assemblies. Acoustics based sensing implemented three sound-emitting devices such as spark gaps mounted on ultrasound transducer, and an array of microphones mounted above the patient in fix position in order to receive the sound pulses from the moving transducer (Fig. 3.17).

The position of recorded 2D images can be calculated based on the knowledge of speed of sound in air, the received sound pulses and the location of microphones. However, the drawback of this protocol is that it may be easily influenced by sound interference from its surrounding environment. The speed of sound might change due to the air humidity, which in turn affecting the calculated geometric parameters. Mechanical arm or articulated arm fixture on ultrasound transducer and movement control by multiple-mechanical joint is also possible to record the relative position and orientation of successive 2D images. But it is limited by its joint orientation, which lower its feasibility on irregular scanning plane (Fig. 3.18).

The general idea of optical sensing protocol and electromagnetic sensing protocol are quite similar, both methods require sensor-marker or -receiver to reflect

Fig. 3.17 Schematic
diagram illustrates the
acoustics based sensing for
position and orientation
tracking

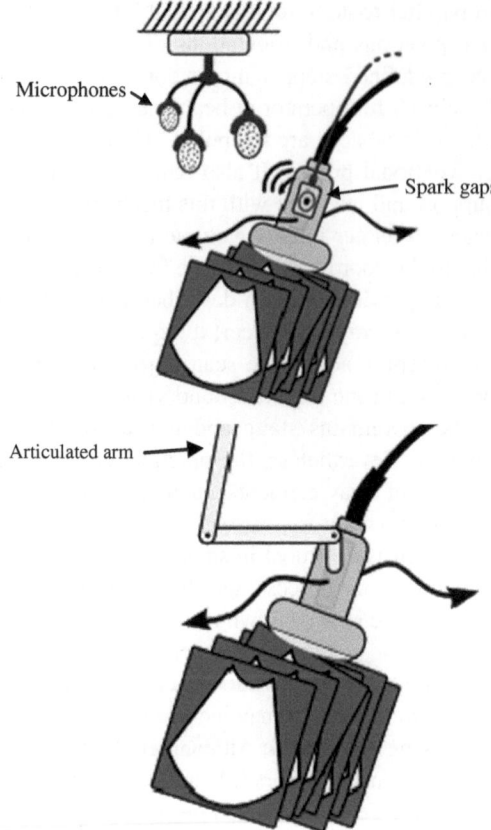

Fig. 3.18 Schematic
diagram showing external
mechanic arm fixture on
ultrasound transducer

the relative geometric space information of the attached US transducer in world
coordinates, either using infrared light reflectors on the cameras, or electromagnetic transmitter. Yet, feasibilities of both methods are limited by its surrounding
environment constraints which could compromise the tracking accuracy. For optical based sensing, it is for sure that the operator must maintain a clear line of sight
between the cameras and the passive/active marker, thus, inconvenient in clinical environment (Gee et al. 2006). In addition, the cameras have limited viewing
volume, distance between the transducer and camera sensor is strictly controlled
in certain scale. The electromagnetic based sensing encounter with the interference from sources such as power cables, electrical signal, medical equipment,
monitors which may influent the local magnetic fields. Any conductive metallic

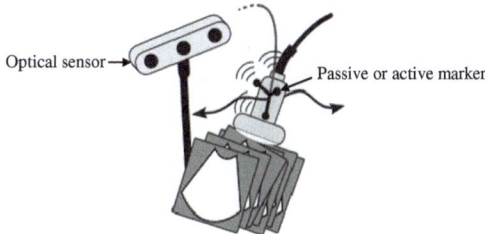

Fig. 3.19 Schematic diagram for optical sensing based position tracking

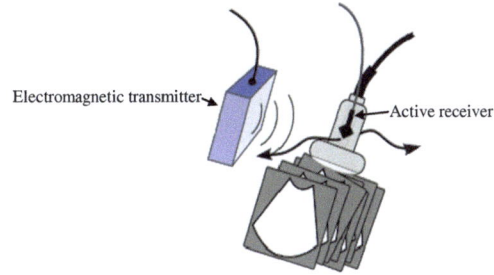

Fig. 3.20 Schematic diagram for electromagnetic sensing based position tracking

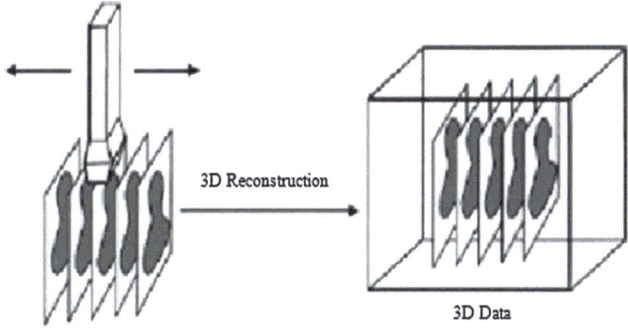

Fig. 3.21 Research protocol of freehand 3D US scanning without tracking devices

ferrous located nearby could also compromise the geometric tracking information (Figs. 3.19 and 3.20).

Due to the lack of viable flexibility and environment constraints, sensor-less freehand 3D ultrasound system are proposed in the present research, and it has become an active research area recently. This freehand protocol does not constraint the acquisition volume, and it gives full clinician freedom to scan interest arbitrary planes without any position sensing devices, and handling the probes over the patient while acquiring successive 2D images, and then reconstruct it into 3D using the fact that pre-known geometric information and the anatomical knowledge

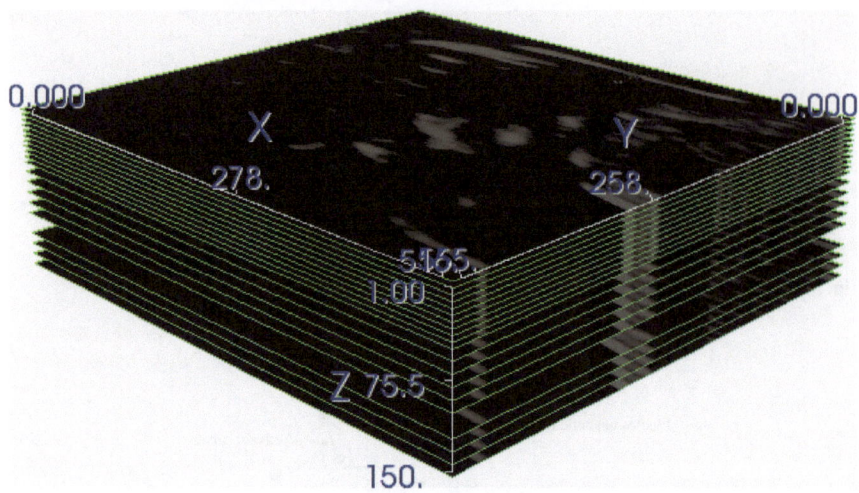

Fig. 3.22 Acquired image embedded in image volume corresponding to its relative order

of interest physiological marker—nuchal translucency. The sonographer must handle the transducer at a regular motion with uniform speed over the small scale of NT scanning area, as shown in Fig. 3.21.

The present freehand 3D US reconstruction processes to generate the anatomy of fetal NT are accomplished in two steps: first, the reconstruction refers to place the acquired 2D US images into 3D formation based on their correct relative positions order. Each image pixel are positioned at its respective 3D coordinates (x, y, z) based on 2D coordinates (x_i, y_i) of that pixel in its 2D image, and the position order (z_k) with respect to 3D coordinates axes. Second, the voxels formation of tetrahedron cube in a regular Cartesian grid of volume elements in three dimensions, which generating 3D representation of fetal anatomy for NT inspection.

As the majority of US imaging is 2D, yet the anatomy of interest marker NT is 3D, therefore, we have investigated the extent of voxel conversion from two dimensional image pixels to comprehend three dimensions NT volume spacing, composited with our rendering algorithm based on lighting; shading model to display the final images. This is essentially important to analyze that the measurement on 3D volumetric possesses higher impact and accuracy vis-à-vis 2D image measurement (Fig. 3.22).

Voxel is the basic unit in three-dimensional image reconstruction where it is analogous to pixel info in conventional 2D images. Voxel uses the defined points to coordinate the direction of gradually increasing order, as shown in Fig. 3.23. The voxels values can be calculated in various interpolation methods dedicated for freehand 3D reconstruction, for example, it may traverse each pixel of the input images and assign its value to one or more voxels of the target grid, or, traverse each voxel of the target grid and collect the necessary information from the 2D input images (Fenster and Downey 2000). For mechanical control scanned images

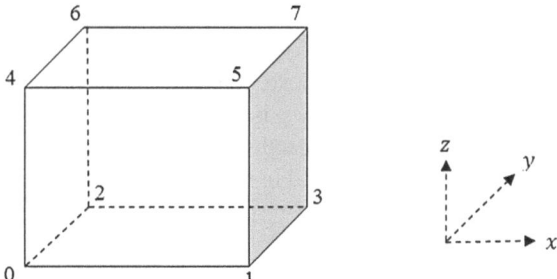

Fig. 3.23 Definition of voxel in three dimensional

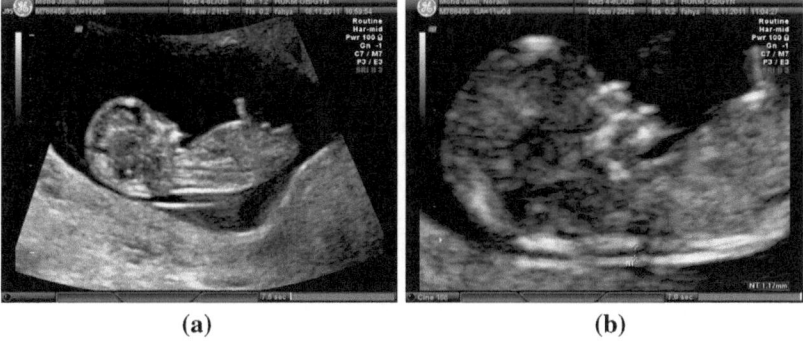

(a)   (b)

Fig. 3.24 Ultrasonography at identical image resolution but different scan depth resulting dissimilar pixel extends: (**a**) Scan depth = 2, (**b**) Scan depth = 5

reconstruction (Tong et al. 1996), normally the interpolation weight are precompiled and placed in a look-up table. In present studies, each pixel is assigned to their respective single voxel without Hole-Filling Step (HFS) because each pixel will not intersect more than one regular voxel array. Otherwise, Pixel Nearest Neighbour (PNN) approach is needed to average the final voxel's values while several pixels may contribute to the same voxel. There are some other interpolation methods includes tri-linear, radial basis function interpolation consider not just the nearest pixel but also the nearby pixels promoting higher quality at the expense of computational speed. However, some experts (Gee et al. 2004; Rohling et al. 1999) have proved that PNN is acceptable especially only for ultrasound imaging modality as compared to other imaging modalities, while other interpolation approaches over PNN interpolation is marginal.

In the present ultrasound image studies, the acquired data's ordinary x and y axis resolution is 800 × 600 with 8 bit depth; it has been resampled and cropped to smaller size to redundant the additional information including patient personal tag-information (Fig. 3.24). One should note that the layout of ultrasound image is different for each subject scanned and it is changes within the scan depths setting, therefore, extraction boundaries have to be calibrated for each data before 3D reconstruction. As the data collected without position tracking devices, we decided to define the voxel spacing in identical values equal to 1. In order to calculate the resampled voxel extend in millimetre scale, we need to adjust the new pixel extend based on their scan depths. This property is crucial as two ultrasound images with identical resolution but different scan depths, the pixel extend between the images are different. The resampled voxel extend in mm scaled can be calculated as follows;

$$Resampled\ Spacing = \frac{S}{N-1} \tag{3.3}$$

where S denotes the actual image size in mm scale; N denotes the number of pixels along S direction.

3.4.2 3D Volumetric Rendering

The notion of volume rendering is to assign each individual voxel element with its color and opacity, by considering its light transmission, emission and reflection for every single voxels (Managuli et al. 2009; Thomas 1994). Transmission of light depends on the opacity of voxels; where light emission depends on the objectness, the greater degree of substance, the stronger the emitted light; and lastly, light reflection depends on the angle of surface and the voxel. In principle, volume rendering steps can be divided into projection, sampling, rendering and compositing. It can handle three-dimensional regular data and irregular data field of volume rendering. The basic idea of the proposed algorithm is based on the ray casting approach which transforms 3D volume data visualization on 2D computer screen by constructing a crossed scalable platform for 3D fetal volumetric formation. This technique based on the visual imagery to construct an ideal physical visual model where each voxel can be seen as the transmission, emission and reflection of light particles. Based on its lightness model and the pixel properties like color, opacity, and the direction along the line of

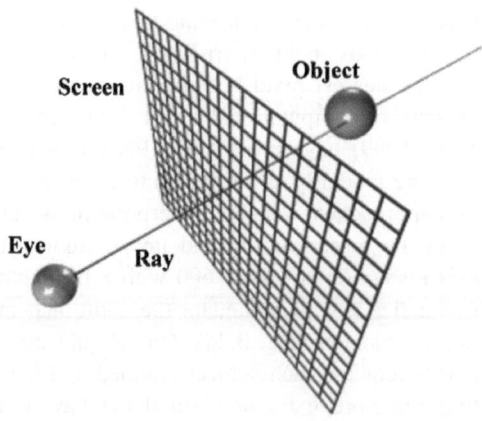

Fig. 3.25 Example of Ray Casting projection

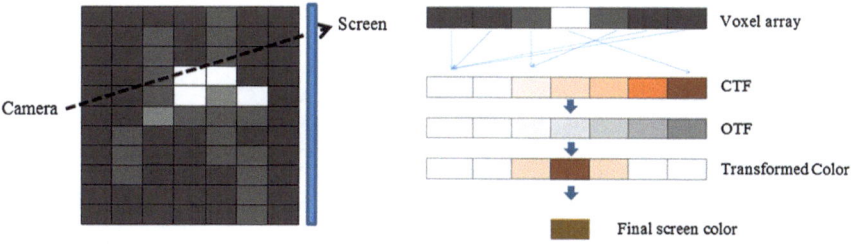

Fig. 3.26 Color composition of a voxel array

observation sight, it forms translucent images on the 3D modeling. Figure 3.25 illustrates the example of ray casting for voxel's array projecting in computing.

The final output value on the computer screen visualization is determined by integrating along the ray's path and by accumulating the grayscale values of the volume by matching them to a composite output color scheme, or called composite transfer function. In present studies, the composite functions are the transformed combination of color transfer function and opacity transfer function. The accumulations of 3D voxels array along the ray are integrated in these two functions to yield the final color visualization on computer screen. The composite ray-based equation is implemented recursively as follows;

$$C'_i = C_i + (1 - A_i)C'_{i-1} \tag{3.4}$$

where C'_i and C'_{i-1} are the computed color at voxel i and $i - 1$ position, C_i is the color emitted at position i, A_i is the operator for opacity absorption at position i. Given a 3D grayscale input array with grayscale value ranges from 0 to N, it is transformed into colored object with increased visualized opacity information as compared to its original data, as expressed below;

$$I \begin{pmatrix} x \\ y \\ z \end{pmatrix}^N \rightarrow I_{RGB} \begin{pmatrix} R_x & G_x & B_x & A_x \\ R_y & G_y & B_y & A_y \\ R_z & G_z & B_z & A_z \end{pmatrix} \tag{3.5}$$

Figure 3.26 illustrates the steps of composition for an input volume array along the ray's path in computing. For each voxel that gets hit by a ray, a new value is added to the array for composite color transformation.

Color transfer function (CTF) is to define the mapping of voxels to an RGB color value. In computing, this function allows for the specification of the midpoint as a normalize distance between nodes, where the midpoint refers to the region that function reaches the average of two bounding nodes; $I \in [0, \ldots, 2^{x-1}]$ and $0 \leq R, G, B \leq 255$. Since all voxels are rendered opaque by default, therefore, we can introduce an opacity transfer function (OTF) to classify the voxels and control their opacity with regard to their scalar value. Similar to CTF, the OTF removes the empty voxels (i.e. background image), and displays varying opacity for the remaining voxels (i.e. organ or tissue of scanned subject). In our case,

(a)   **(b)**

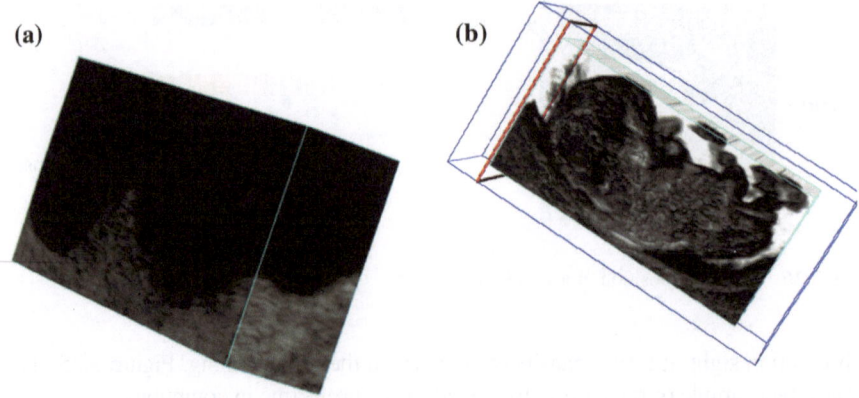

Fig. 3.27 volumetric rendering of 3D ultrasound data. **a** Existing commercial rendering scene before composition rendering. **b** After composition rendering

these transfer functions are often attached to this attribute and used to define more specifically the appearance of the volume properties, thus makes the NT marker more apparently. Figure 3.27 shows part of the simulation results of volumetric rendering using reconstructed 3D ultrasound volume data.

Although a complete anatomical survey was more frequently referred to real-time 2D images; the examination of 3D volumes, irrespective of fetal position, provided details information for an inclusive review of the fetal anatomy and NT fold thickness measurement. If analyzed image planes were not in appropriate position, the anatomical marker will not be clearly shown, or it might be measured longer or shorter than its real length. Therefore, it was crucial to identify and display the best 2D image plane from 3D model. Volumetric rendering is achieved by using algorithms to vary opacity, transparency, and depth in projecting the volume information onto a single image. Fetal head, face, neck and several ultrasound markers could be examined using 3D volumes within suitable composite integration of OTF and CTF. These structures represented the main sites in which fetal abnormalities can be seen in the first trimester. By referring to 2D images, the anatomy of these structures could be examined more frequently, but the difference compared with 3D volumes was not statistically significant. An advantage of 3D volume reconstruction is that obtained volumes can be manipulated; by rotating the image to a standard orientation, such as longitudinal, transverse, and coronal, desired views can be achieved allowing the 3D architecture of these structures to be examined in detail. This is of central importance for the measurement of fetal NT thickness in opting the best sagittal plane by 3D rather than 2D images. A fine-tuning 3D model would be achieved by suitable color and opacity setting; this aids the identification of the location of NT in fetus. Real-time 2D ultrasound is yet the gold standard in examining fetal anatomy nowadays. However, this pilot research shows that 3D volumes can provide maximum information in a very limited reconstruction time. It also offers a new way of storing and conveying information. 3D volumes in digital format can be sent and examined by a specialist, allowing the execution of a virtual scan instead of rescanning the patient. As a result, the pivot in 3D imaging include acquisition time and resolution, have improved the solution of current limitations and enhanced feasibility of 3D volume's routine practice in early pregnancy.

3.4.2.1 Virtual Slider Extraction Designs

In order to investigate the internal structure of nuchal translucency explicitly, we have proposed a three dimensional virtual slider cutting plane. Qualitative analysis of the proposed experimental three-dimensional nuchal translucency visualization system can be achieved by developed virtual slider in 360° freedom orientation. Medical doctors can generate any angle plane with interactive operations through slide, rotate and tilt of the cutting plane at any combination of x, y or z position. This function is prominent to the existing commercial MPR visualization techniques which is limited in conventional view modes includes sagittal, coronal and axial plane. The cutting plane will be implemented as a grid plane with an arrow direction, which indicates the current position of three dimensional grid planes. The movement of virtual slider can be controlled through the interaction of mouse click and movement triggered. The geometric rotation matrix transformation of the 3D graphics can be expressed as follows;

$$T_{3D} = \begin{bmatrix} a_{11} & a_{12} & a_{13} & a_{14} \\ a_{21} & a_{22} & a_{23} & a_{24} \\ a_{31} & a_{32} & a_{33} & a_{34} \\ a_{41} & a_{42} & a_{43} & a_{44} \end{bmatrix} \tag{3.6}$$

where $\begin{bmatrix} a_{11} & a_{12} & a_{13} \\ a_{21} & a_{22} & a_{23} \\ a_{31} & a_{32} & a_{33} \end{bmatrix}$ Performing geometric rotation through the changing of

nine parameters; determine the cutting plane direction. Whenever the reconstructed 3D models rotate about the x axis, the coordinate vertex of y and z change while x is not. Given a rotated angle θ_x, as shown in Fig. 3.28; the transformation of T_{3D}^x is as below;

Fig. 3.28 Three dimensional rotation transformation matric

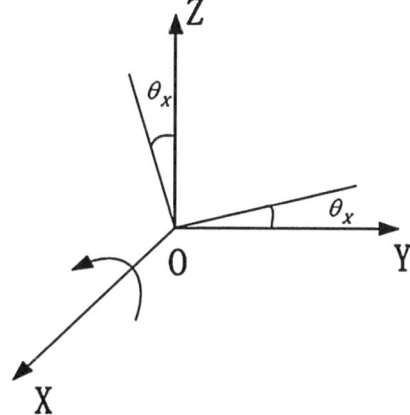

$$T_{3D}^x = \begin{bmatrix} 1 & 0 & 0 & 0 \\ 0 & \cos\theta_x & -\sin\theta_x & 0 \\ 0 & \sin\theta_x & \cos\theta_x & 0 \\ 0 & 0 & 0 & 1 \end{bmatrix} \qquad (3.7)$$

Similarly, the rotation transformation matric about the axis y and z are expressed as follows;

$$T_{3D}^y = \begin{bmatrix} \cos\theta_y & 0 & \sin\theta_y & 0 \\ 0 & 1 & 0 & 0 \\ -\sin\theta_y & 0 & \cos\theta_y & 0 \\ 0 & 0 & 0 & 1 \end{bmatrix} \qquad (3.8)$$

and

$$T_{3D}^z = \begin{bmatrix} \cos\theta_z & -\sin\theta_z & 0 & 0 \\ \sin\theta_z & \cos\theta_z & 0 & 0 \\ 0 & 0 & 1 & 0 \\ 0 & 0 & 0 & 1 \end{bmatrix} \qquad (3.9)$$

To perform this, the algorithms need an external library; *AddObserver* on developed plane class, which specifies the interception event such as *vtkCommand::InteractionEvent*. Whenever the position of the slider is changed, the location information need to be updated, and activating the changing of the rendering view. This interactive simulation requires a *Callback* command to answer the mouse call event. The parent class of this *Callback* command is *vtkCommand*; it will obtain the latest location information by adding its member function *PlaneWidget*, which acts as a pointer to record the new cutting plane location. Figures 3.29 and 3.30 illustrates the function of virtual slider which acts as orientation free cutting plane in various positions on reconstructed 3D ultrasound fetal training phantom and clinical fetal images respectively. The entire 3D model may be arbitrarily rotated in any direction to obtain the optimum NT ultrasonic marker view, in this way; the operator always has the accurate 3D image-based cues relating the plane that being manipulated to the fetal anatomy.

3.2 VTK Widget Event Handling Mechanism

VTK Widget can be defined as the geometry and behavior control of the displayed object information. It allows the direct interaction of programmer with the data in three-dimensional data field manipulation. The widget control is depending on the mouse click and move triggered. It will receive the activated control of interactive events and generate appropriate behaviors according to the sign given by the users. The widget features are separated into two parts, one part inherited from *vtkAbstractWidget* class for event handling, and the other part inherited from *vtkWidgetRepresentation* class for geometry description, as shown in Fig. 3.31. It can be observed that *vtkWidgetRepresentation* is combines with *vtkAbstractWidget* subclass to produce a 3D widget.

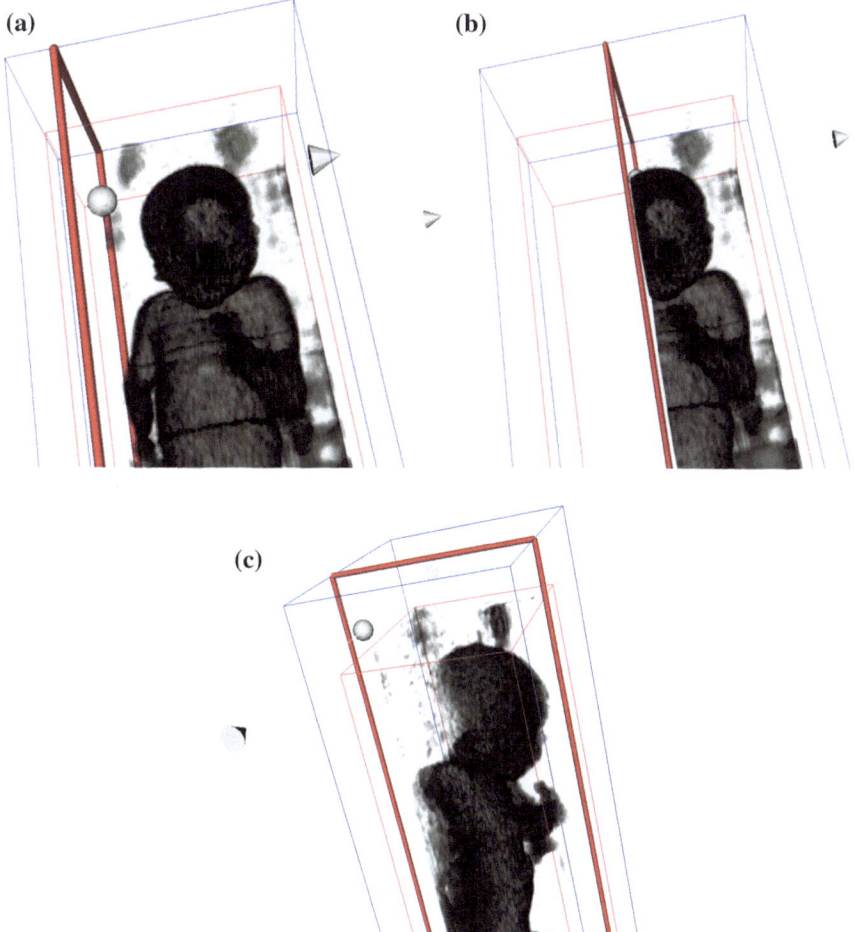

Fig. 3.29 Simulation results of slider on ultrasound fetal training phantom: (**a**), (**b**) and (**c**) Various virtual slider position

Although the quality of the reconstructed 3D fetal images depends critically on the composed data's fidelity, the viewing technique used to display the 3D findings often play a dominant role in determining the information transmitted to the operator, therefore, the benefit of the proposed pipeline mechanism visualization system is that it enables parallelized processing for various hybrid visualization modes designs. It includes our proposed synchrony view of 3D volume with 2D plane extraction, hybrid MPR integrated with volumetric rendering, and expansion of virtual slider in 2 or 6 planes extraction. For synchrony view, we design to create two renderer pointers for each form of data in parallel visualized on the same renderer window, both pipeline channel's output are responded to the same window interactor that reflecting the updated position and orientation. The important of synchrony view in the present studies is to allocate the position of re-slice 2D plane that contains ultrasonic marker NT within the movement control of virtual slider's geometric parameters (Fig. 3.32).

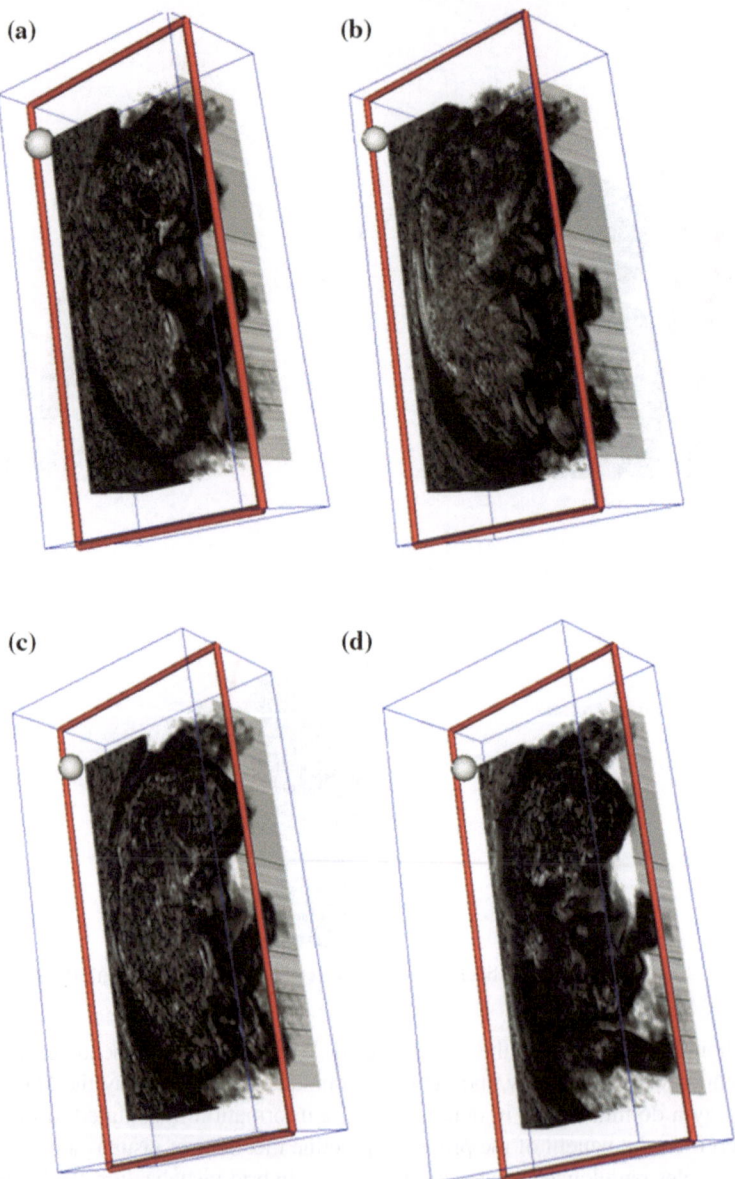

Fig. 3.30 Various position of slider on clinical NT profile scanning: (**a**), (**b**), (**c**) and (**d**) Various virtual slider position

As the operator changes the position and orientation of the red-frame slider as shown in Fig. 3.33, it will extract the particular 2D plane in real-time based on the pre-reconstructed 3D coordinates and matric transformation. Figure 3.34 illustrates another simulation results on ultrasound fetal NT profile in arbitrary plane extraction. Figure 3.35 illustrates the similar hybrid visualization as described

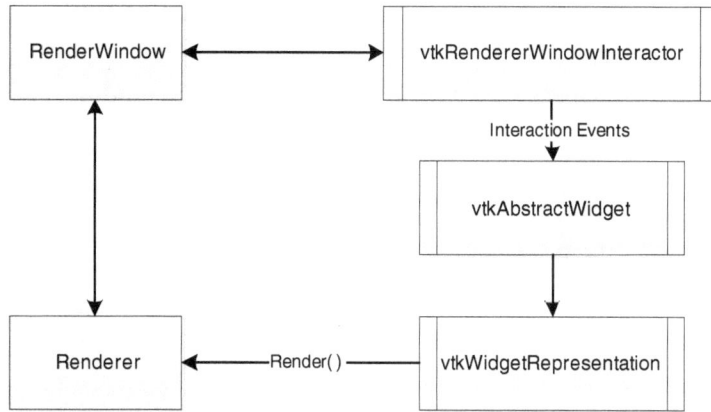

Fig. 3.31 VTK Widget event handling description

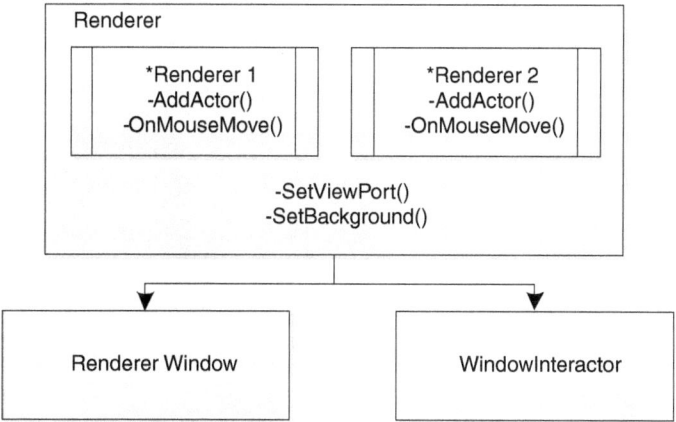

Fig. 3.32 Parallel renderer pipeline connection in hybrid visualization

above. The proposed design and implementation encounters the existing limitation of 2D prenatal scanning protocol where the healthcare professional judged
the selected 2D plane (with ultrasonic markers) using hand-eye coordination and
imagination of fetal structure based on their experiences. It improves the efficiency
of plane selection for marker investigation.

```
int x=30;
Plane2->SetNormal(-Plane->GetNormal()[0],-Plane->GetNormal()[1],-Plane->GetNormal()[2]);
Plane2->SetOrigin(Plane->GetOrigin()[0]+x*Plane->GetNormal()[0],
                  Plane->GetOrigin()[1]+x*Plane->GetNormal()[1],
                  Plane->GetOrigin()[2]+x*Plane->GetNormal()[2]);
vMapper->AddClippingPlane(Plane2);

Plane->SetOrigin(Plane->GetOrigin()[0]-x*Plane->GetNormal()[0],
                 Plane->GetOrigin()[1]-x*Plane->GetNormal()[1],
                 Plane->GetOrigin()[2]-x*Plane->GetNormal()[2]);
```

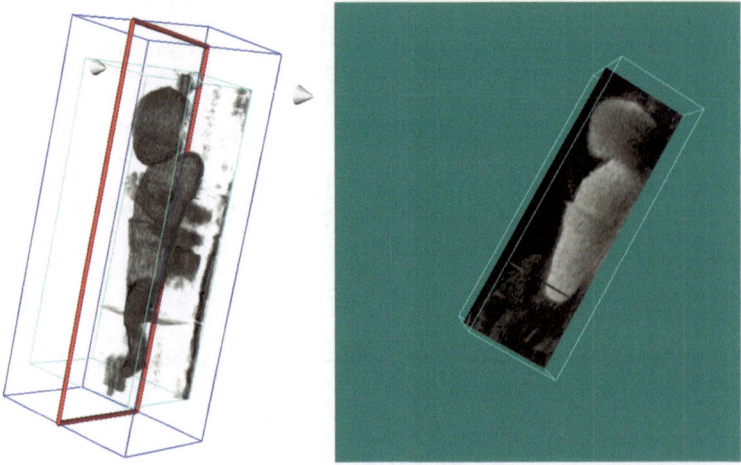

Fig. 3.33 Simulation of proposed synchrony view for ultrasound fetal training phantom 2D arbitrary plane extraction based on real-time slider control

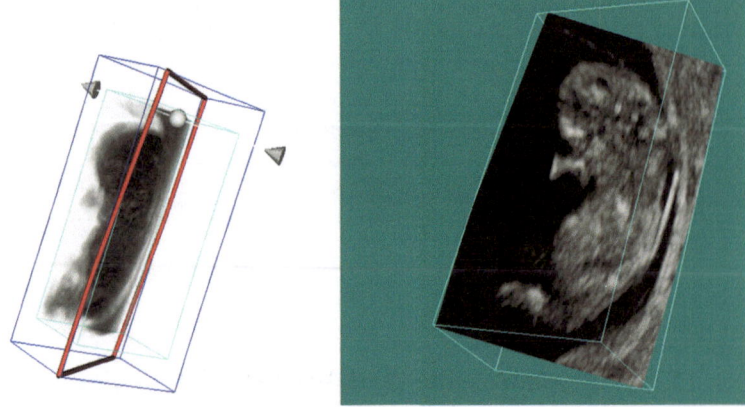

Fig. 3.34 Simulation of synchrony view for re-slice 2D clinical NT data extraction based on real-time orientation virtual slider arbitrary plane control

As a single slider has the sign character with positive and negative on each rendering side to hinder or liberate the rendering scene respectively, we can also design the double sided cutter by adding an extra slider in opposite sign direction with a positive integer spacing x in between the planes. The computing realization is shown as follows;

Figure 3.36 shows the simulation results of double sided cutter on reconstructed 3D ultrasound images over various sizes spacing between the cutting planes. This visualization technique is essential useful for mid-range NT marker region assessment. This similar technique can be extended to six cutting planes which established a cutting boxes formation, as shown in Fig. 3.37.

Fig. 3.35 Hybrid MPR integrated volumetric view mode on, **a** ultrasound fetal training phantom, and **b** clinical NT data: (**a**) Simulation using ultrasound fetal phantom, (**b**) Simulation using collected clinical data

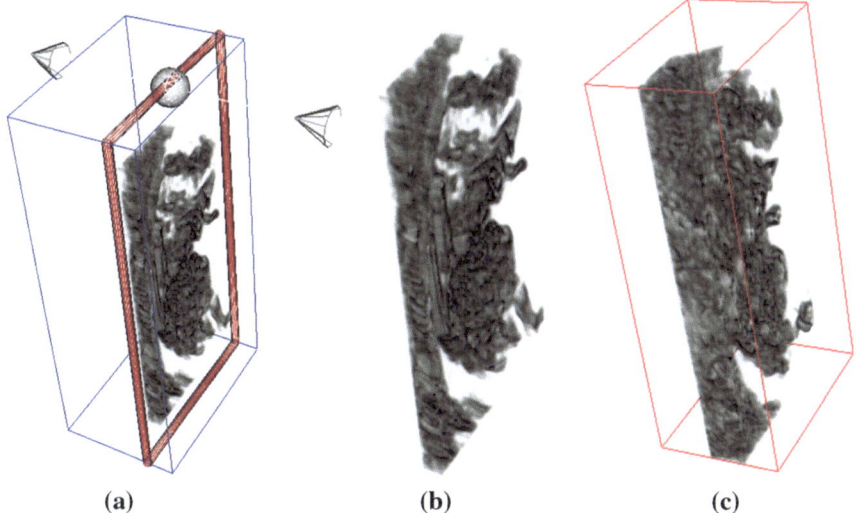

Fig. 3.36 Simulation of double sided cutter using 2 planes slider **a b** *x* spacing equal to 15 pixels **c** *x* spacing equal to 30 pixels

3.5 Three Dimensional Nuchal Translucency Segmentation

Ultrasound medical imaging is widely used nowadays in clinician application occasioned by its intuitive, convenient, safety, non-invasiveness, and low cost consumption (Laporte and Arbel 2010; Min and Myoung 2007; Lee et al. 2007). Nevertheless, the restrictions on the imaging mechanism lead to low resolution in ultrasound images and they have been always a major drawback in its imaging modality. This condition will be more complicated when the subjects of interest are the inhomogeneity fine structure of organ or tissue. Some minor structures

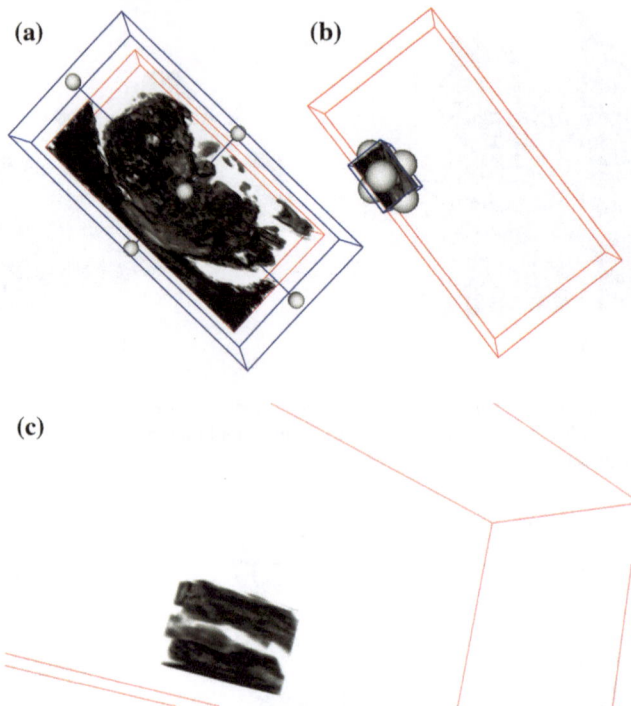

Fig. 3.37 Simulation of box-shape cutter formation using 6 planes slider (**a**) 3D reconstructed fetal images within the box formation (**b**) (**c**) respective ROI of cutting planes residual

cannot be resolved by ultrasound formation, and when it is coupled with the interference of acoustic signals, the ultrasound images form a well-known unique spot, namely speckle noise. Obviously, it will greatly reduce the ultrasound image quality, and make the segmentation and analysis of image detail—NT structure become more challenging.

A wide broad of segmentation techniques have been considered based on the properties of ultrasound fetal characteristics and NT properties. There are four common segmentation techniques and widely applied in image processing; threshold based, boundaries based, region based and hybrid based segmentation techniques. Threshold based segmentation are based on the postulate that pixels grey value lies to certain set of class. It does not cope well with the blur edge images noise and neglect all other spatial information except those belonging to the defined class. This method assumes that pixels within an object contain a range of grey levels. Let g represents an image and $g(x, y)$ represents gray-level intensity for pixels on coordinates (x, y) in the image; T represents thresholding level that manipulate thresholding segmentation, the thresholding process can be represented mathematically as followings:

$$g(x, y) = 1, \quad \begin{array}{l} \text{If } g(x, y) < T \\ \text{Otherwise } g(x, y) = 0 \end{array} \tag{3.10}$$

The above equations show binary thresholding. Another type of thresholding method is multilevel thresholding which represented by followings:

$$g(x, y) = label\ 1, \quad \text{If } g(x, y) < T_1 \tag{3.11}$$

$$g(x, y) = label\ 2, \quad \text{If } T_1 < g(x, y) < T_2 \tag{3.12}$$

$$g(x, y) = label\ 3, \quad \text{If } g(x, y) > T_2 \tag{3.13}$$

Other types of thresholding method include local or adaptive thresholding. This type of thresholding method determines the thresholding level T by considering the locations of the pixels in the image as opposed to global thresholding which uses a single threshold over the entire image.

The weakness of thresholding method is that the determination of T is not a trivial task. Typically, it needs human visual system to determine the threshold or it needs to be determined through empirical results. The thresholding level can also be determined from the image's information, for instance, the statistical value of the image like mean and standard deviation. Another frequently implemented determination of T procedure is done via through the inspection of the image's histogram: the valleys points of gray level in the histogram are used as thresholding level. Uneven illumination of the image often makes the automated thresholding level searching fail or become inaccurate especially while using the global thresholding. Besides, the thresholding method does not consider the relationship among pixels and also the existence of noises could seriously affect the resultant image after segmented.

Boundaries based segmentation techniques perform greatly in image with abrupt edge changes; in other words, large mean difference of neighboring pixel intensities. This method can implement the classic gradient operator such as Canny, Sobel or Roberts filter. This method bases on the concept that an object in an image will have discontinuity in gray level to other object or background, or in other words, bases on the concept that a change of grey level intensity across the objects border. The boundary based segmentation basically is the extension of conventional edge detection in 2D digital image using gradient operator. The extension of detected edges is called the edge linking process.

After the edges of the objects are detected using the gradient operator, these edges will be linked together to form a close loop and fulfill the definition of segmentation: partition of objects from the images. The linkages are formed based on information obtained during the process of generating the edge pixels: the strength of the edge pixels shown in (3.14) and the vector direction shown in (3.15) of the edge pixels.

$$\nabla f = \sqrt{\left(\frac{\partial f}{\partial x}\right)^2 + \left(\frac{\partial f}{\partial y}\right)^2} \tag{3.14}$$

$$\beta(x, y) = \tan^{-1}\left(\frac{\partial f}{\partial y} \bigg/ \frac{\partial f}{\partial x}\right) \tag{3.15}$$

Let (x_1, y_1) be the coordinates of the edge pixels (x_2, y_2) be the coordinates of the neighborhood pixels around the edge pixels an T_1 and T_2 denotes nonnegative integer gradient strength threshold and vector direction threshold. The below criteria based on gradient strength and direction need to be satisfied in order to consider whether (x_2, y_2) can be linked to (x_1, y_1) as an elongated edge:

$$|\nabla f(x_1, y_1) - \nabla f(x_2, y_2)| < T_1 \qquad (3.16)$$

$$|\beta(x_1, y_1) - \beta(x_2, y_2)| < T_2$$
$$\qquad (3.17)$$

If both criteria of (3.16) and (3.17) are achieved, the edge pixels neighborhood pixels are considered similar with the edge pixels and they are linked. This process is repeated all over the image. The weakness of this method is the unrealistic assumption that the detected edges will form a continuous edges line surrounding the objects. Firstly, in practical implementation, the edges detected hardly found located exactly on the boundary of the objects due to uneven illumination or other artifacts of the images like noises.

The properties of NT structures are the formation of fold thickness resulting two echogenic lines: distal and proximal layer. By understanding the ultrasound image formation, the accumulation of fluids in between the layers will form homogeneity of ultrasound pixels information; visualized as dark tone. Relying on both facts above: ultrasound speckle noise and NT structural, region based segmentation integrated with seeded growing technique appears prominent as compared to others attributable to the single interest region is the central homogeneous dark regions in between the fold thickness. Therefore, we proposed an optimized 3D seeded region growing (SRG) integrated into our overall methodological design model in order to segment the interest ultrasound marker; NT uses a pre-processing 3D diffusion method.

3.5.1 Pre-Processing 3D Diffusion

The preprocessing of 3D diffusion aims to protect the edge of NT boundary while smoothing and de-noising the image features without distortion. In recent years, filtering method based on the diffusion equation emerges as a popular filtering technique for ultrasound medical image. It is based on the principal in solving the initial value of input image using nonlinear heat diffusion equation. Perona and Malik (1990) initiate the proposed anisotropic diffusion equation (referred as P-M model, referring Eqs. 2.8, 2.9 and 2.10) for image denoising. Their diffusion model has aroused great interest among the researcher worldwide in the last decades, causing a boom on partial differential equation enhancement in image processing. By introducing image features, diffusion equations are designed to control the appropriate spread of the diffusion coefficient, producing the smooth image and enhancing the image feature information

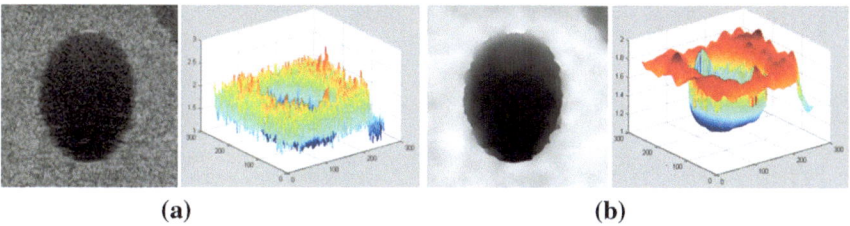

Fig. 3.38 Diffusion based on number of iteration. **a** Diffusion iteration = 1. **b** Diffusion iteration = 200

simultaneously. However, the direct practicing of P-M model on medical ultrasonic image is not satisfactory. The inefficiency of P-M model is predicated on its inherent incapability in solving the additive noise. For ultrasonic images with additive noise, P-M model shows promising de-noising effect. Nevertheless, the resulting outcome of P-M model on multiplicative noise or speckle in ultrasound image shows very limited effect, sometimes even counterproductive. Over a decade of research and exploration on anisotropic diffusion enhancement, it is now a powerful speckle noise removal called SRAD for 2D ultrasound images (Yu and Acton 2002). The SRAD exploits the instantaneous coefficient of variation (ICOV), which is shown to be a function of the local gradient magnitude and Laplacian operators. The diffusion coefficient in 2D SRAD is a function of q, where $q(x, y, t)$ are shown as follows;

$$c(x, y; t) = \frac{1}{1 + \frac{[q^2(x,y;t) - q_0^2(t)]}{q_0^2(t)} \left(1 + q_0^2(t)\right)} \tag{3.18}$$

$$q(x, y; t) = \sqrt{\frac{\left(\frac{1}{2}\right)\left(\frac{|\nabla I|}{I}\right) - \left(\frac{1}{4^2}\right)\left(\frac{\nabla^2 I}{I}\right)^2}{\left[1 + \left(\frac{1}{4}\right)\left(\frac{\nabla^2 I}{I}\right)\right]^2}} \tag{3.19}$$

And the initial speckle scale computed homogenous region q_0 is shown as follows;

$$q_0(t) = \sqrt{\frac{\mathrm{var}[z(t)]}{z(t)}} \tag{3.20}$$

where I is input image, ∇ is gradient operator, | denotes the magnitude. The function shows high value at edge and low value in homogenous region. Moreover, the diffusion order can be revealed from our simulations in Fig. 3.38 which demonstrates that ICOV is successfully smoothened the speckled region and intensified the contrast between constructive object and noise area.

Value of diffusion coefficient is influenced by ICOV where edge pixel area activates a lower diffusion coefficient and indirectly decreases the level of diffusion. Speckle area produces higher value of diffusion coefficient which increases

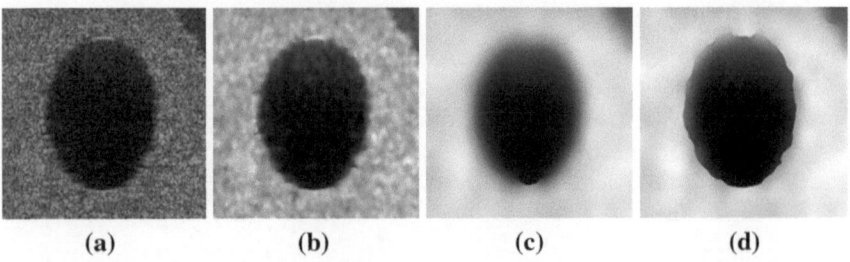

(a)   (b)   (c)   (d)

Fig. 3.39 Level of smoothing. **a** Original ultrasonic image. **b** Under-smoothing. **c** Over-smoothing. **d** Smoothing at appropriate iteration

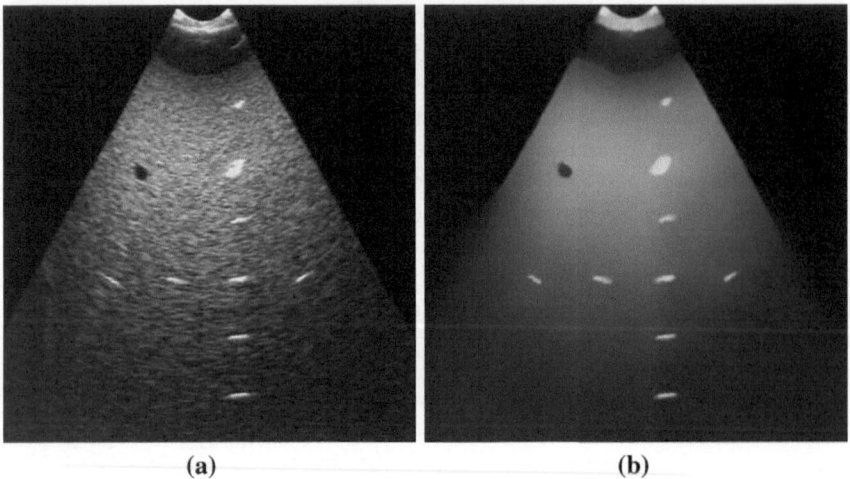

(a)   (b)

Fig. 3.40 Simulation results comparison using ultrasound training phantom. **a** Before 2D SRAD. **b** After 2D SRAD

the smoothing effect in certain area. Figure 3.39 illustrates the smoothing of ultrasound image with difference number of iteration and smoothing time step.

Ultrasound image changes gradually in every iteration. Larger iteration will generate a smoother image but it requires longer processing period and over-smoothing may destroy the edge of ultrasound image. Figure 3.40 shows our simulation results on ultrasound phantom with appropriate diffusion iteration.

Much of the research attentions are focus on 2D diffusion implementation, rather than 3D ultrasound images. Theoretically, conventional 2D SRAD diffusion equations are also able to diffuse in 3D ultrasound images by performing diffusion on each slice of collective 2D images before reconstructing them into 3D model. Nonetheless, these render an obstructive effect in space direction of the 3D model in response to the missing information between slices of 2D image. This simple method will results an obstructive effect in 3D volume, especially in

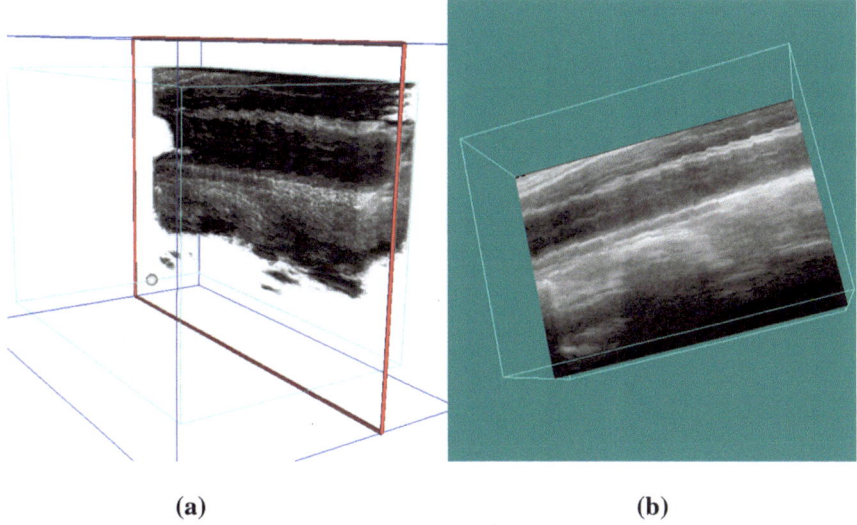

(a)   (b)

Fig. 3.41 Simulation results in 3D ultrasound volume. **a** XZ view on 3D visualization. **b** 2D re-slice of viewed XZ plane

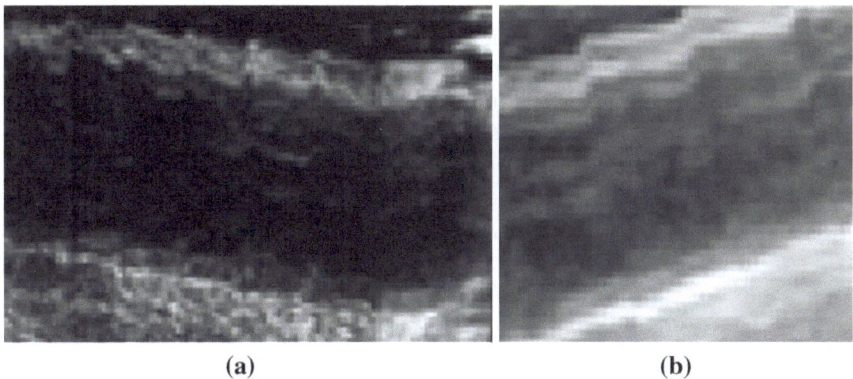

(a)   (b)

Fig. 3.42 Obstructive line effect. **a** On 3D volume rendering. **b** On 2D ultrasonic slices

Z-pixels direction due to the missing information in spacing track. Image features in between neighboring slices were not conserved during implementation of 2D SRAD. In our investigation, we have simulated this limitation on a series of 2D ultrasound images, collected on a volunteer's carotid artery at BMTI, Ilmenau, as shown in Fig. 3.41.

Close observation on Fig. 3.41 can found that obstructive effects are present in spacing direction, as shown in Fig. 3.42. Measurement on XY direction will not encounter with this limitation, but this effect will causing wrong measurement

Fig. 3.43 3D simulation
result of ultrasound
fetal training phantom
reconstruction and
visualization before 3D
diffusion

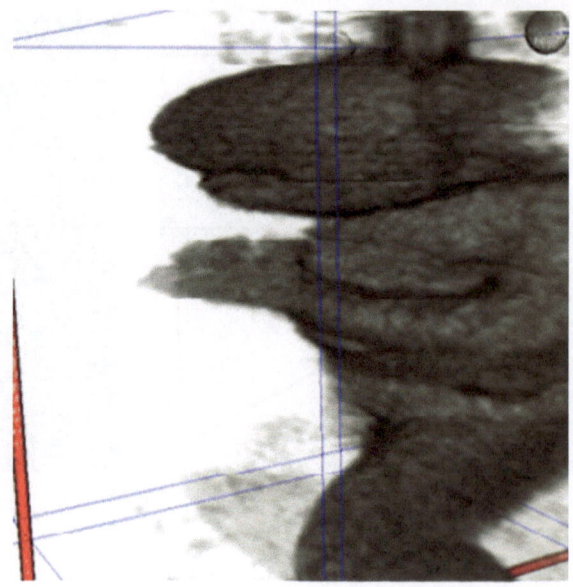

especially when the subject of interest are in small scale. Important morphology
was destroying when we visualize in 3D volume mode.

   3D diffusion algorithms are integrated into pre-processing in present studies.
It will operate directly on voxels in 3D volumetric images rather than every pixel
in 2D slices. Thus, information between slices of 2D image are explored and preserved. Suppose the $I(x, y, z, t)$ denotes the input image at t stage in the continuous
domain, where ∇I denotes image gradient, t depicts the time parameter. The 3D
diffusion coefficient can be expressed as below;

$$\frac{\partial}{\partial x} I(x,y,z,t) = div[g(x,y,z,t) * \nabla I(x,y,z,t)] \qquad (3.21)$$

where,

$$g_1(\|\nabla I\|) = exp\left(-\left(\frac{\|\nabla I(x,y,z,t)\|}{k}\right)^2\right) \qquad (3.22)$$

$$g_2(\|\nabla I\|) = \frac{1}{1 + \left(\frac{\|\nabla I(x,y,z,t)\|}{k}\right)^{1+\alpha}}, \alpha > 0 \qquad (3.23)$$

where k is a constant, set for adjusting the 'definition of edge'. This value is normally determined by the noise level of the image and the intensity of the edges
in image. It is significant for diffusion function to recognize the edges and thus
diffusion operation is diminished on them. With the intent to smooth the surface
of the NT structure and facilitate the subsequent processing of segmentation,

Fig. 3.44 3D simulation
results of ultrasound
fetal training phantom
reconstruction and
visualization after 3D
diffusion

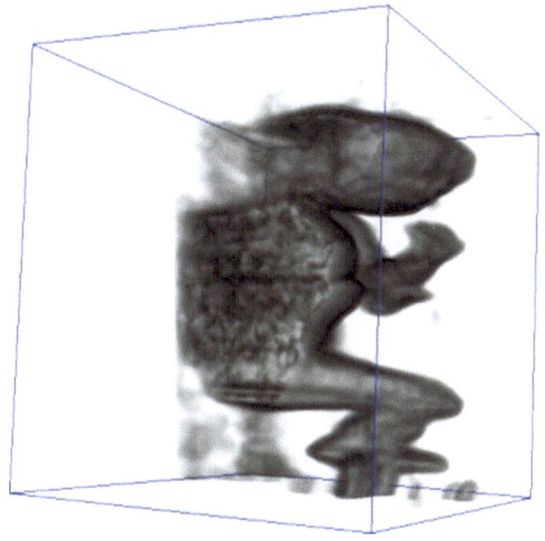

especially segmentation involves stopping criteria for region based growing; the
image underwent anisotropic diffusion with the iterative update on each voxel in
the image by the flow intensity contributed by its neighboring voxels. In present
studies, we have tested the proposed algorithms on fetal training phantom and
carotid artery before the implementation on clinical NT profile. Figures 3.43, 3.44,
3.45, 3.46, 3.47, 3.48, 3.49 and 3.50 show the present simulation results for 3D
reconstruction and visualization of ultrasound fetal training phantom, volunteered
carotid artery and clinical NT scanned profile respectively; before and after 3D
diffusion.

Distinguished differences of Figs. 3.43 and 3.44 are the homogeneous areas
been smoothed, meanwhile the sharp boundaries showing clear edge on the fetal
skin are pertained. Besides, we have also tested the algorithm on scanned volunteered carotid arteries. Figure 3.45 shows internal view of interest of carotid
arteries before 3D diffusion. It is a real time free orientation 3D visualization
in C++ platform on 2.4 GHz Pentium 4 processing unit. Within appropriate
empirical variables, the internal view of carotid arteries can be examined through
virtual 3D slicers in any orientation. Similar to 2D ultrasound slices; it can be
observed that the restructured 3D volumetric model using un-processed 2D ultrasound slices comes with speckle noise. Besides, the internal structures of carotid
arteries are also contaminated by the background cloud, which makes the challenges on the following processes of precise segmentation and measurement in
small scale assessment.

Fortunately, both difficulties are resolved by the proposed diffusion in 3D as
shown in Fig. 3.46. We can observe that whole volume assembly was smoothed
and the background clouds were eliminated. For example, Fig. 3.47 shows the
close views comparison sliced out from Figs. 3.45b and 3.46b.

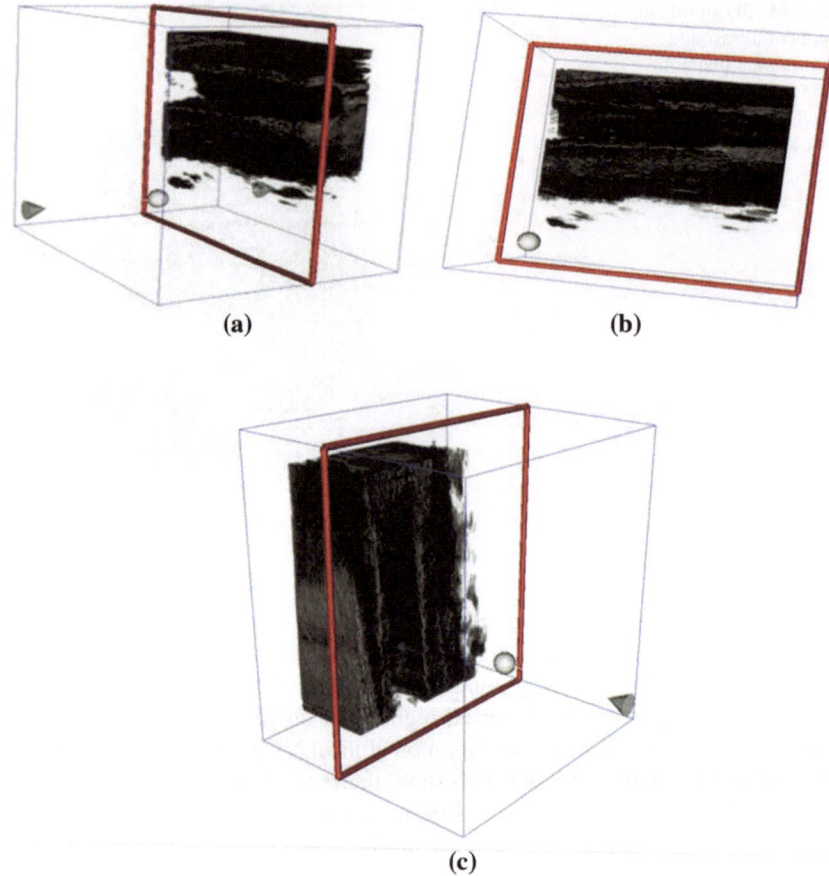

Fig. 3.45 3D simulation results of volunteered carotid artery visualization before 3D diffusion: (**a**), (**b**) and (**c**) Various virtual slider and rendering position

To inspect the qualitative performance of proposed diffusion method in clear way, we have performed a detail comparison assessment on same 3D volume at the same position. Figures 3.48, 3.49, 3.50 shows the cloud-noises were eliminated effectively after 3D diffusion.

3.5.2 Three Dimensional Seeded Region-Based Segmentation

In three dimensional imaging systems, medical computed tomography will produce digital image data which covered internal body structures, such as Nuchal Translucency. In order to analyze the internal object interest in terms of their shape, form and its function, biological extraction from the restructured

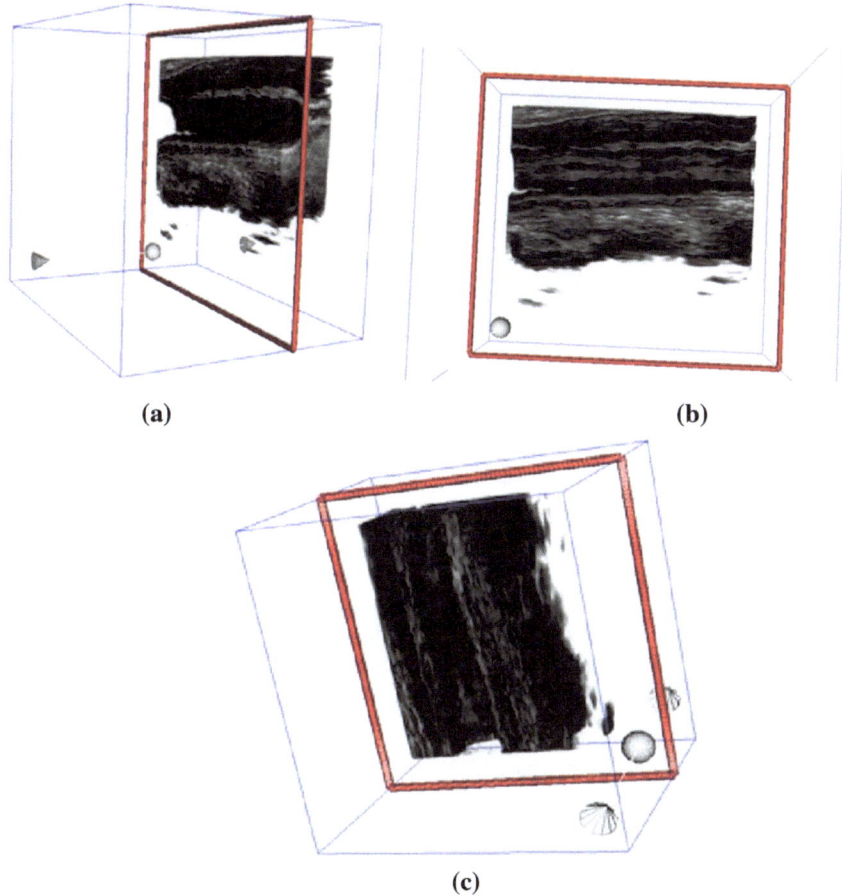

(a)   (b)

(c)

Fig. 3.46 3D simulation results of volunteered carotid artery visualization after 3D diffusion: (**a**), (**b**) and (**c**) Various virtual slider and rendering position

visualization is necessary. The process of these extractions is called medical image segmentation. The proposed technique in current ultrasound marker segmentation is seeded region based approaches integrated into our 3D reconstruction and visualization system. The fundamental of algorithm designs is a collection of similar pixels in 3D nature together to form the interest regions; fold thickness of NT. Let T be the set of all pixels before being allocated. The elements of the set T are (x, y, z). R_i denotes the regions, it could be formed by a single seed or a group of pixels. n denotes the total seeds. $N(x, y, z)$ denotes the 26 immediate neighborhood pixels of each element (x, y, z) of set T (Fig. 3.51).

$$T = \left\{ (x, y, z) \notin \bigcup_{i=1}^{n} R_i \,\middle|\, N[(x, y, z)] \cap \bigcup_{i=1}^{n} R_i \neq \phi \right\} \qquad (3.24)$$

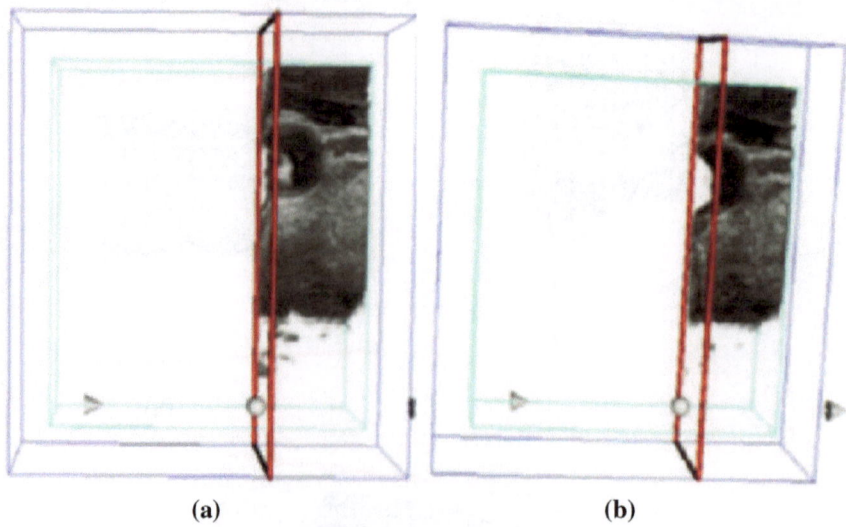

(a)	**(b)**

Fig. 3.47 Simulation of carotid artery, **a** before 3D diffusion and, **b** after 3D diffusion

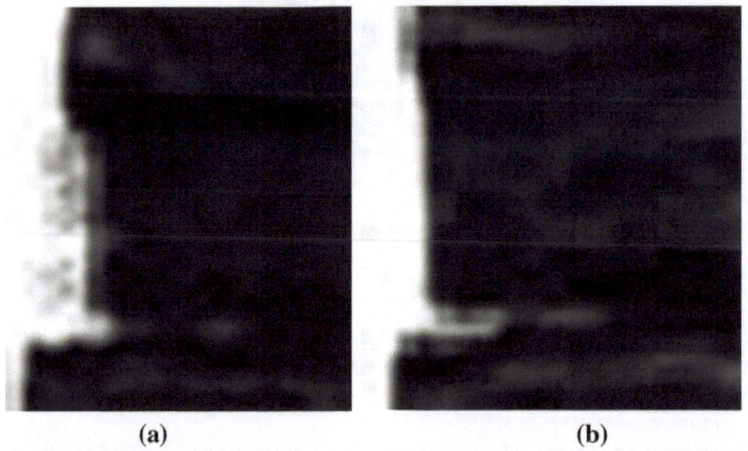

(a)	**(b)**

Fig. 3.48 Comparison of simulation results at 3D edge **a** before diffusion, and **b** after diffusion

Equation (3.24) describes that being a neighborhood pixel of a region, they must fulfill two properties: (1) being unallocated (2) overlapping with at least one pixel with the region R_i. Determining whether a pixel belongs to a region is typically done by comparing the pixel and region relationship in terms of difference or similarity. Normally, a similarity test is to be carried out using a metric. To establish a quantitative measurement of degree of homogeneity to the region, a metric has to be defined:

$$H[T(x,y,z)] = |T(x,y,z) - mean(R_i)| \qquad (3.25)$$

Fig. 3.49   Real Patient 3D
fetal reconstructions before
3D Diffusion

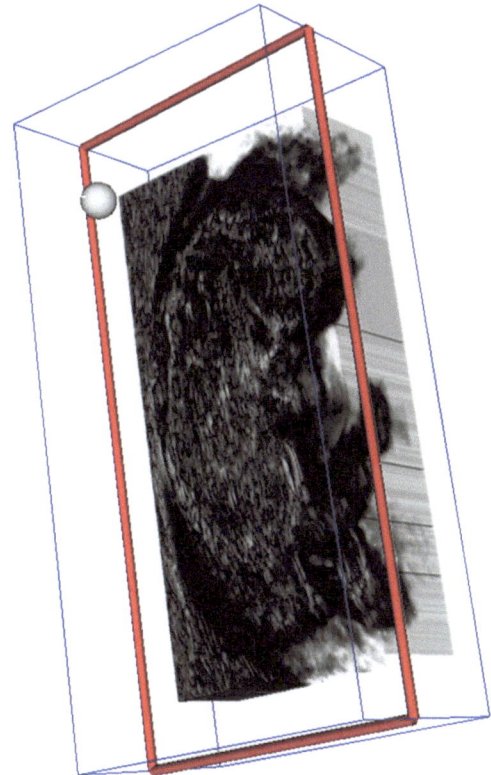

Fig. 3.50   Real Patient 3D
fetal reconstructions after 3D
Diffusion

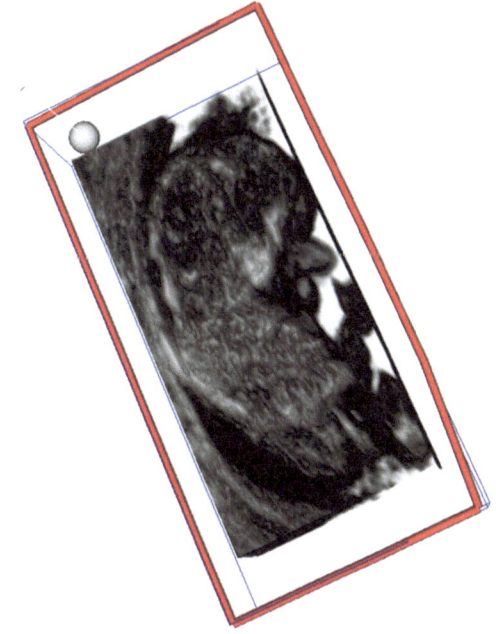

Fig. 3.51 Computational
model designs for 3D SGR of
NT segmentation

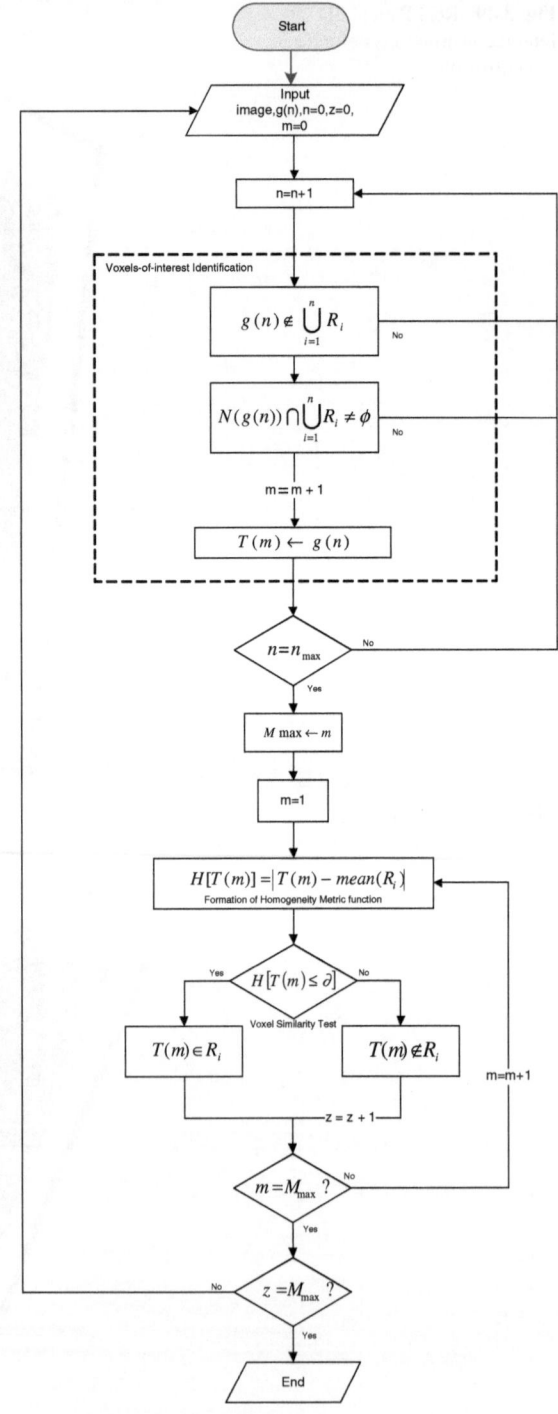

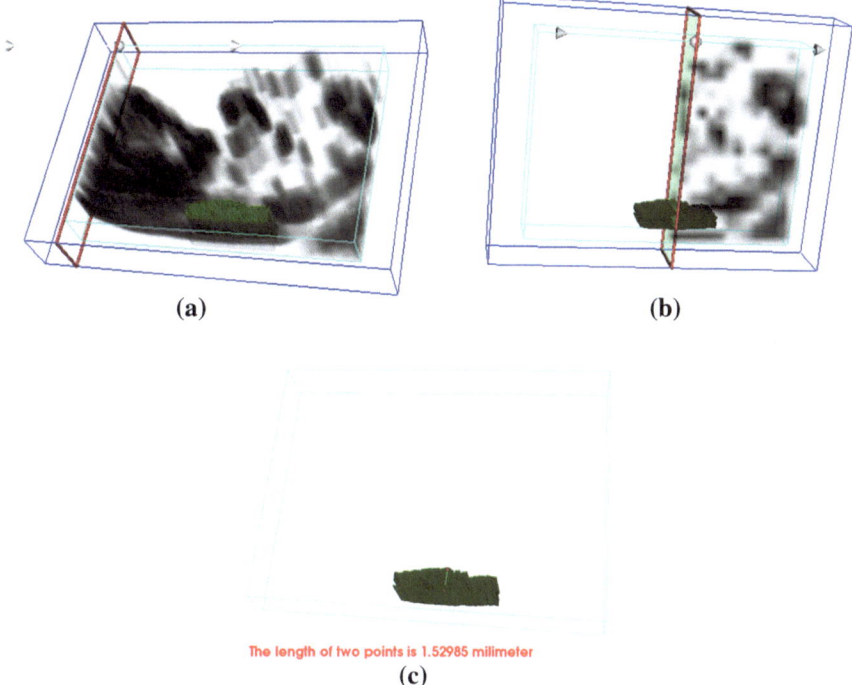

(a)   (b)

(c)

Fig. 3.52 Three dimensional nuchal translucency simulation. **a** 3D volumetric rendering. **b** Slicing view to show for NT segmentation. **c** Thickness measurement using 3D Euclidean approach

$H(.)$ denotes the metric function for each element in set T with the purpose of calculating the quantitative degree of homogeneity between the target neighbor pixels with the region's expected value. The rule-based determination of region membership:

$$T\ (x, y, z) \in R_i, \quad \text{If } H[T(n) \le \partial]$$

(3.26)

∂ can be chosen via empirical result or using an automated threshold setting technique. Equation (3.26) describes that if the homogeneity metric $H\ |T(x, y, z)|$ is less than ∂, then it would be considered as one of the member element of $R_i(x, y, z)$. If $T(x, y, z)$ are situated as immediate neighbor pixels of m regions, then the pixel will be allocated to region that yields smaller $H\ |T(x, y, z)|$ which can be represented mathematically as followings:

$$T(x, y, z) \in \text{Min } R_i(x, y, z) \quad \text{For } i = 1 : m, m \ge 1$$

(3.27)

It is claimed that the most effective operation is by semi-interactive where human use their own interpretation of the suitable location for seed placing. As the human operator knows the number of seeds in various specific application and the accurate location of seed placing to avoid any anomalies being chosen as seed in automated region growing. The flowchart below illustrates the computation of the

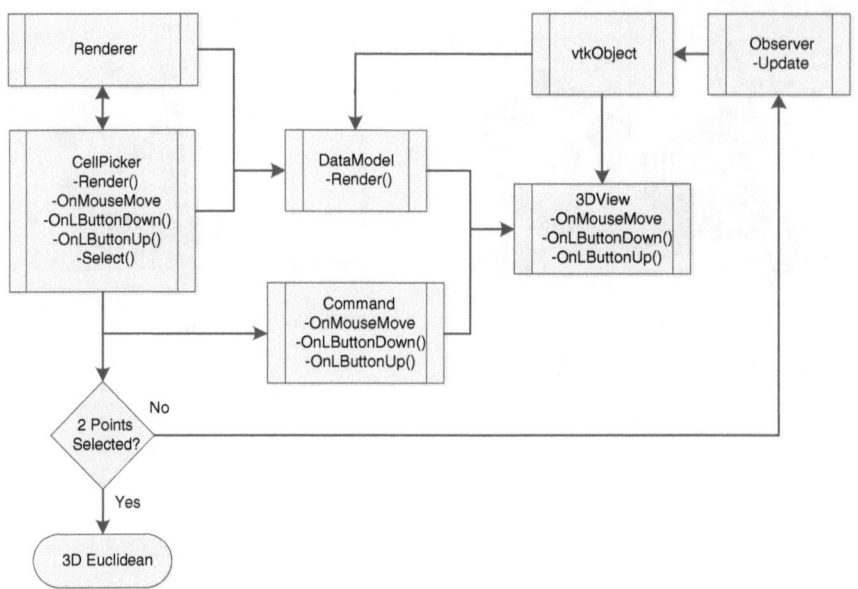

Fig. 3.53   Algorithm connection between *CellPicker* and renderer data channel for 3D Euclidean computation

abovementioned SRG algorithm. Figure 3.52 shows the simulation results of 3D NT segmentation. It is followed by interactive measurement control as explained in the following section.

3.5.3   Three Dimensional Nuchal Translucency In Vivo Measurement

As described in Sect. 3.4.2.2 for widget event handling mechanism, the final NT rendering scene can be divided into two models, operator 3D widget model and renderer data model. In order to update the control information of operator 3D widget model on renderer data model, *Callback* command is required to activate the response accordingly. We have proposed to apply the *AddObserver* to act as an external module in order to perceive and reflect the real time data model status. An important advantage of 3D US imaging is the ability to measure the volume length in arbitrary geometric orientations. While a qualitative visual assessment is often valuable, quantitative measurement also provide a more reliable basis for decision making and diagnostics as compared to B-mode US scans. In present research, computing three dimensional measurement of NT is accomplished through *CellPicker* widget. *Observer* will convert the *CellPicker* data into geometrical and physical parameters in the rendering scene. Once *CellPicker* updates its parameters, *Observer* will update the data simultaneously through the interface

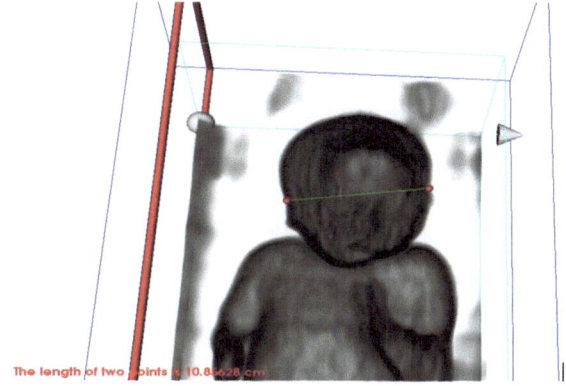

Fig. 3.54 Example simulation of 3D measurement on reconstructed ultrasound fetal training phantom

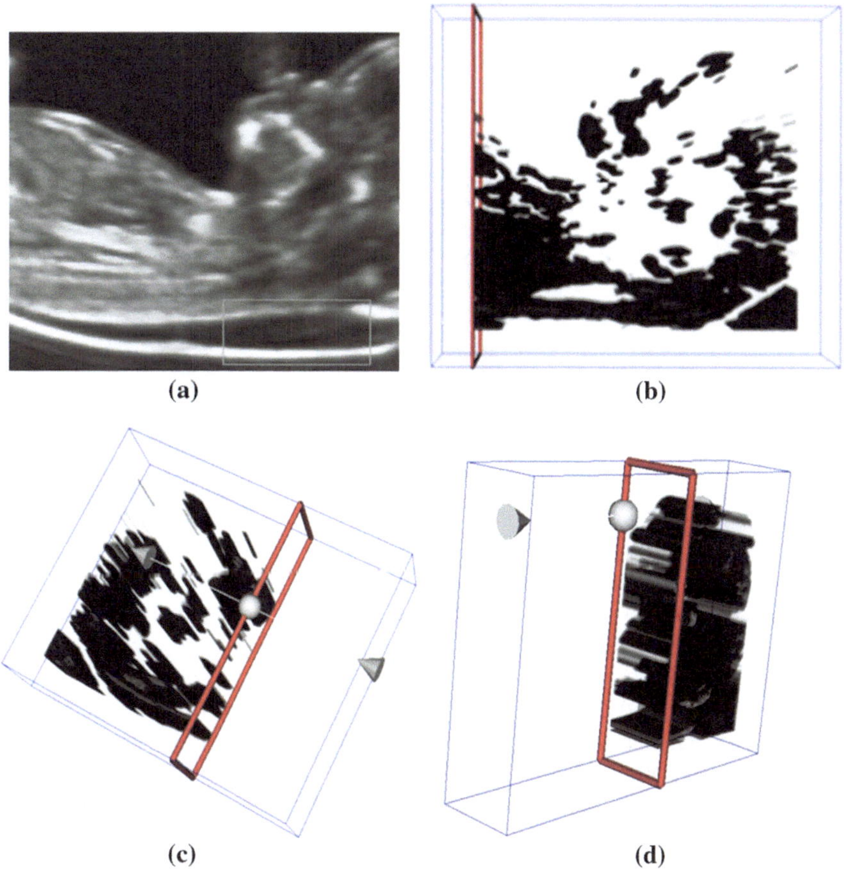

Fig. 3.55 Ultrasound marker Nuchal translucency measurement (**a**) 2D ultrasound image (**b**), (**c**) and (**d**) Simulated 3D ultrasound image for 3D fold thickness

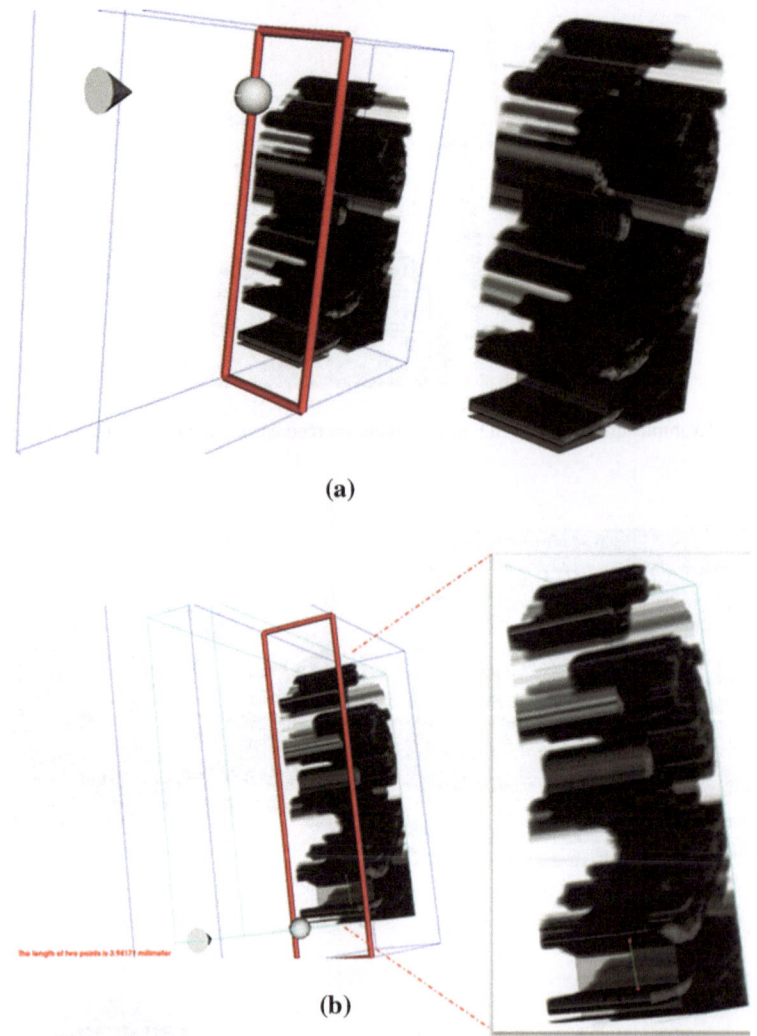

Fig. 3.56 Prominent view of internal structure ultrasound marker. **a** Changed position of virtual slider and zoom visualization for clear ultrasound marker viewing. **b** Three dimensional NT marker measurements length

connection established. Figure 3.53 shows the interactive algorithm connection for three dimensional measurement of nuchal translucency.

The proposed system will send mouse and keyboard input messages to *CellPicker* through command updating. For the three dimensional Euclidean distance computation, two worlds Cartesian coordinate are required. The point picker in computer screen is only two dimensional coordinates which has no direct relation with the three dimensional objects, and therefore, conversion into world coordinates is necessary. The distinct difference between two dimensional coordinates and world coordinates is that the latter contains depth information, which reflects the depth of Z-coordinate spaces. In current studies, the selected two dimensional coordinates will be introduced as a straight line extended into Z spacing, which might penetrate the three dimensional rendering objects followed by acquiring the coordinate intersection with the surface of rendering objects. Straight line may intersect with more than one plane; therefore, several interactions are possible.

The algorithms compare the distance of all the intersecting point with the initial point coordinates. The coordinates with the shortest distance calculated will be selected as the three dimensional world coordinates. To compute the NT folds thickness; let's assume the selected two points are $P(x, y, z)$ and $Q(x, y, z)$. The vector of PQ or QP is formed by two 3D coordinates; $E(\overline{pq})$ can be calculated using Euclidean approach as follows;

$$E(\overline{pq}) = \sqrt{\sum_{i=1}^{n}(q_i - p_i)^2} \qquad (3.28)$$

where E denotes the Euclidean distances, i denotes the dimension of vectors; in our case; $n = 3$. Figure 3.54 illustrates the 3D measurement on the reconstructed ultrasound fetal training phantom, the red sphere represents the geometric formation of selected two coordinates, where the green line indicates the vector formed for 3D distance computation.

Figure 3.55 shows the simulation results of clinical NT measurement in 3D form as compare to conventional 2D ultrasonic images. The measurement is taken placed on prominent surface of 3D NT structure, rather than pixel selection on 2 echogenic lines, therefore, overestimation or underestimation of fold thickness will be resolved while encountered the optimum plane selection (Fig. 3.56).

References

Abuhamad, A. (2005). Technical aspects of nuchal translucency measurement. *Seminars in Perinatology, 29*(6), 376–379.

Altmann, K., Shen, Z. Q., Boxt, L. M., King, D. L., Gersony, W. M., Allan, L. D., et al. (1997). Comparison of three-dimensional echocardiographic assessment of volume, mass, and function in children with functionally single left ventricles with two-dimensional echocardiography and magnetic resonance imaging. *American Journal of Cardiology, 80*(8), 1060–1065.

Bekker, M. N., Twisk, J. W. R., & van Vugt, J. M. G. (2004). Reproducibility of the fetal nasal bone length measurement. *Journal of Ultrasound in Medicine, 23*(12), 1613–1618.

Berg, S., Torp, H., Martens, D., Steen, E., Samstad, S., Hoivik, I., et al. (1999). Dynamic three-dimensional freehand echocardiography using raw digital ultrasound data. *Ultrasound in Medicine and Biology, 25*(5), 745–753.

Chen, M., Yang, X., Wang, H. F., Leung, T. Y., Borenstein, M., Nicolaides, K., et al. (2010). Learning curve in measurement of fetal frontomaxillary facial angle at 11–13 weeks of gestation. *Ultrasound in Obstetrics and Gynecology, 35*(5), 530–534.

Fenster, A., & Downey, D. B. (1996). 3-D ultrasound imaging: A review. *IEEE Engineering in Medicine and Biology Magazine, 15*(6), 41–51.

Fenster, A., & Downey, D. B. (2000). Three-dimensional ultrasound. *Imaging, 2*, 457–475.

Thomas, F. (1994). Raycasting of nonregularly structured volume data. *Computer Graphics Forum, 13*(3), 293–303.

Gee, A., Prager, R., Treece, G., & Berman, L. (2003). Engineering a freehand 3d ultrasound system. *Pattern Recognition Letters, 24*(4–5), 757–777.

Gee, A., Prager, R., Treece, G., Cash, C., & Berman, L. (2004). Processing and visualizing three-dimensional ultrasound data. *British Journal of Radiology, 77*, S186–S193.

Gee, A. H., Housden, R. J., Hassenpflug, P., Treece, G. M., & Prager, R. W. (2006). Sensorless freehand 3d ultrasound in real tissue: speckle decorrelation without fully developed speckle. *Medical Image Analysis, 10*(2), 137–149.

Gobbi, D. G., & Peters, T. M. (2002). Interactive intra-operative 3d ultrasound reconstruction and visualization. *Medical Image Computing and Computer-Assisted Intervention - MICCAI 2002. 5th International Conference. Proceedings, Part II (Lecture Notes in Computer Science vol. 2489)*: 156-163163.

Gobbi, D. G., Comeau, R. M., & Peters, T. M. (1999). Ultrasound probe tracking for real-time ultrasound/MRI overlay and visualization of brain shift. In C. Taylor & A. Colchester (Eds.), *Medical image computing and computer-assisted intervention, Miccai'99, Proceedings* (Vol. 1679, pp. 920–927). Berlin: Springer.

Gobbi, D. G., Comeau, R. M., & Peters, T. M. (2000). Ultrasound/MRI overlay with image warping for neurosurgery. In S. Delp, A. M. DiGioia, & B. Jaramaz (Eds.), *Medical image computing and computer-assisted intervention—Miccai 2000* (Vol. 1935, pp. 106–114). Berlin: Springer.

Min, K.-J., & Myoung, K.-H. (2007). Image enhancing technique for high-quality visual simulation of fetal ultrasound volumes. *Systems Modeling and Simulation, 337–341*

King, D. L., Gopal, A. S., Keller, A. M., Sapin, P. M., & Schroder, K. M. (1994). 3-Dimensional echocardiography—advances for measurement of ventricular volume and mass. *Hypertension, 23*(1), I172–I179.

Laporte, C., & Arbel, T. (2010). Measurement selection in untracked freehand 3D Ultrasound. *Medical image computing and computer-assisted intervention: MICCAI. International Conference on Medical Image Computing and Computer-Assisted Intervention* (13(Pt 1), pp. 127–134).

Lee, Y.-B., Kim, M.-J., & Kim, M.-H. (2007). Robust Border Enhancement And Detection For Measurement Of Fetal Nuchal Translucency In Ultrasound Images. *Medical & Biological Engineering & Computing, 45*(11), 1143–1152.

Managuli, R., Karadayi, K., Canxing, X., & Yongmin, K. (2009). Volume rendering algorithms for three-dimensional ultrasound imaging: image quality and real-time performance analysis. *IEEE International Ultrasonics Symposium, 2009*, 2324–23272327.

Meairs, S., Beyer, J., & Hennerici, M. (2000). Reconstruction and visualization of irregularly sampled three- and four-dimensional ultrasound data for cerebrovascular applications. *Ultrasound in Medicine and Biology, 26*(2), 263–272.

Nelson, T. R., & Pretorius, D. H. (1998). Three-dimensional ultrasound imaging. *Ultrasound in Medicine and Biology, 24*(9), 1243–1270.

Ohbuchi, R., & Fuchs, H. (1991). Incremental volume rendering algorithm for interactive 3d-ultrasound imaging. *Lecture Notes in Computer Science, 511*, 486–500.

Perona, P., & Malik, J. (1990). Scale-space and edge detection using anisotropic diffusion. *IEEE Transactions on Pattern Analysis and Machine Intelligence, 12*(7), 629–639.

Rohling, R., Gee, A., & Berman, L. (1999). A comparison of freehand three-dimensional ultrasound reconstruction techniques. *Medical Image Analysis, 3*(4), 339–359.

Tong, S., Downey, D. B., Cardinal, H. N., & Fenster, A. (1996). A three-dimensional ultrasound prostate imaging system. *Ultrasound in Medicine and Biology, 22*(6), 735–746.

Trobaugh, J. W., Trobaugh, D. J., & Richard, W. D. (1994). 3-Dimensional imaging with stereotaxic ultrasonography. *Computerized Medical Imaging and Graphics, 18*(5), 315–323.

Wee, L. K., Chai, H. Y., & Supriyanto, E. (2011). Surface rendering of three dimensional ultrasound images using Vtk. *Journal of Scientific and Industrial Research, 70*(6), 421–426.

Yu, Y., & Acton, S. T. (2002). Speckle reducing anisotropic diffusion. *IEEE Transactions on Image Processing, 11*(11), 1260–1270.

Chapter 4
Clinical Tests and Measurements

Abstract Statistical analysis and comparison results between current 2D assessment method and the proposed 3D method are conducted. In this section, we will analyze the degree of association, limits of agreement and repeatability of measurements for both 2D and 3D methods comparison. Intra-observer variability is tested to answer the repeatability skills between operators' measurements. Pearson's correlation value can be computed to evaluate the relationship between existing method and proposed new method. Levels of agreement between methods are analyzed using Bland–Altman's test on three varied grouping observations. Repeatability coefficient is calculated to evaluate the measurements repeatability between the methods. This chapter looks into the test for the measurement results of proposed method. The findings are evaluated with the existing 2D measurement method. It includes the descriptive analyses, degree of association, limits of agreements, Paired sample t-test and repeatability between the methods.

4.1 Introduction

In this section, we will analyze the degree of association, limits of agreement and repeatability of measurements for both 2D and 3D methods comparison. These clinical measurements can be analyzed using different quantification between observations made by using 2D and 3D methods on the same subjects in several ways; the differences between methods, the differences within same methods or so called repeatability, and the variability of inter and intra-observers measurements.

Since the true values of measured clinical data are unknown, in other words, measurement of nuchal translucency (NT) using same method by same operator on same patient in successive times will not be constant, so both methods are compared neither providing an unequivocally true measurement, correlation

K. W. Lai and E. Supriyanto, *Detection of Fetal Abnormalities Based on Three Dimensional Nuchal Translucency*, SpringerBriefs in Applied Sciences and Technology, DOI: 10.1007/978-981-4021-96-8_4, © The Author(s) 2013

and regression analysis alone is not enough and misleading, therefore, assessing the limits of agreement between two methods for clinician measurement is proposed by Bland and Altman (1986, 1995, 1999, 2007). This is because some lack of agreement between different methods of measurement is inevitable. The 95 % limits of agreement can be interpreted by mean difference ±1.96 time standard deviation of the differences between methods. It also can be used to investigate the repeatability within same method on multiple measurements on same patient, and calculating the repeatability coefficient (RC).

The results analyses starts with the normality testing on the collected data using Kolmogorov–Smirnov test. Intra-observer variability is analyzed to exam the repeatability skills of operator for data collection, inter-observer variability is excluded from the data as the clinician work of operator at hospital are not reduplicated on the same individual patients. Next, level of agreement and its association between two methods are computed. It follows by the same analyses on method repeatability analysis.

4.2 Descriptive Analysis

Total numbers of 23 clinician data are collected from Hospital Universiti Kebangsaan Malaysia (HUKM) from August 2011 until February 2012. The measured ultrasonic marker is nuchal translucency, or simply NT using conventional 2D B-mode prenatal ultrasound scan protocol. The type of transducer implemented in current examination is abdominal probe, with beam form 3.5 MHz frequency. The target group of examined patients is pregnant women in first trimester singleton pregnancies. Populations included are Malaysian only, and three successive trials 2D ultrasonic NT measurements for each patient were taken. Ethical Committee (Human—JEPeM) at Hospital Universiti Kebangsaan Malaysia approved all the clinical study materials. The three trials 3D measurements were carried out using our developed 3D reconstruction and measurement program.

The measurements result for 2D and 3D method are listed in Table A in the Appendix C. Figure 4.1 above illustrates the standard deviation of 2D and 3D measurement based on the Table A. The data distribution for collected clinician NT measurement is shown in Fig. 4.2 below. The figure shows the Gaussian bell shape data distribution for conventional 2D ultrasound measurement method. Graphically the curve is described in terms of the point at which its height is maximum (mean) and how wide it is (standard deviation).

Computed mean and standard deviation within-subject for both measurement methods are $\mu_x = 1.4428 \pm s_{xw} = 0.13664$ (2D) and $\mu_y = 1.4084 \pm s_{yw} = 0.05514$ (3D). Note that one extreme data dispersion from the bell shape is the only case from existing data collection with high risk Trisomy 21 syndrome (P20—refer to Table A at Appendix), where the measurement is larger than 2.5 mm [based on guideline from Fetal Medicine Foundation (FMF),

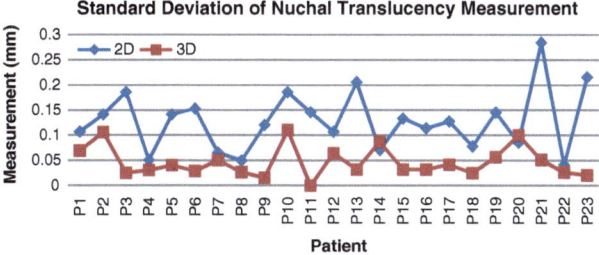

Fig. 4.1 Standard deviation of nuchal translucency measurement for 2D and 3D method respectively

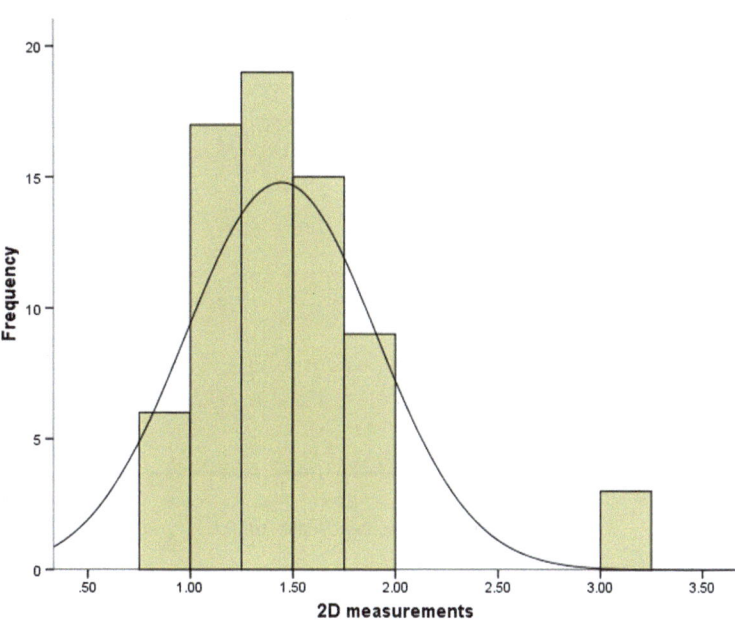

Fig. 4.2 Histogram distributions of 2D ultrasound nuchal translucency thicknesses

UK]. To confirm the normal distribution of the analyses data with parametric symmetry, Kolmogorov–Smirnov test can be used, as shown in Table 4.1 below. Result shows that 2D nuchal translucency ultrasonic markers measurements are normal distributed with significant P values equal to 0.383.

Figure 4.3 below illustrates good correspondence between 3D and 2D measurements. More than 95 % of ultrasonic NT marker measurements fall in the range of 1.0–2.0 mm. Collectively, it reflects the fact that these statistics are within normal ranges based on FMF reports. It is also agreed from the feedback of the collaborated practitioners.

Table 4.1 One-Sample Kolmogorov–Smirnov test

N		69
Normal parameters[a, b]	Mean	1.4428
	Std. deviation	0.46536
	Absolute	0.109
Most extreme differences	Positive	0.109
	Negative	−0.095
Kolmogorov–Smirnov Z		0.907
Asymp. Sig. (2-tailed)		0.383

a Test distribution is normal
b Calculated from data

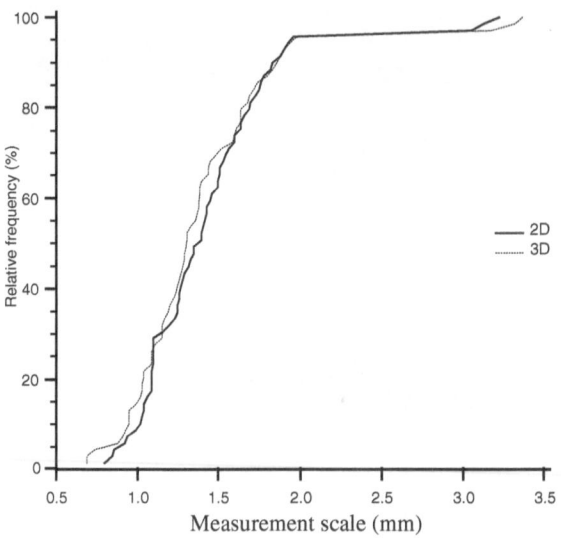

Fig. 4.3 Relative cumulative frequencies of method comparisons

Figure 4.4 shows the scatter plot comparison between 2D and 3D methods, the red dot-line indicates the line of equality; (a) considered two pairs of mean for each method for comparison, so sample size equal to 23, while (b) taking all measurement trials of each method as input, sample size equal to 207.

The Pearson's correlation coefficient values, R for plot (a) is 0.9039 with 95 % confidential interval (CI) from 0.8387–0.9035 and (b) 0.8750 with 95 % CI from 0.7836–0.9588 respectively at significant level $P < 0.0001$, indicates a strong positive linear correlation relationship between the methods, as 2D NT measurement increases, 3D NT measurement increases as well. The computed linear regression line equation between methods for both plots is $y = 0.3043 + 0.8072x$ and $y = -0.05004 + 1.0121x$; where 3D measurements as variable y, 2D measurements as variable x.

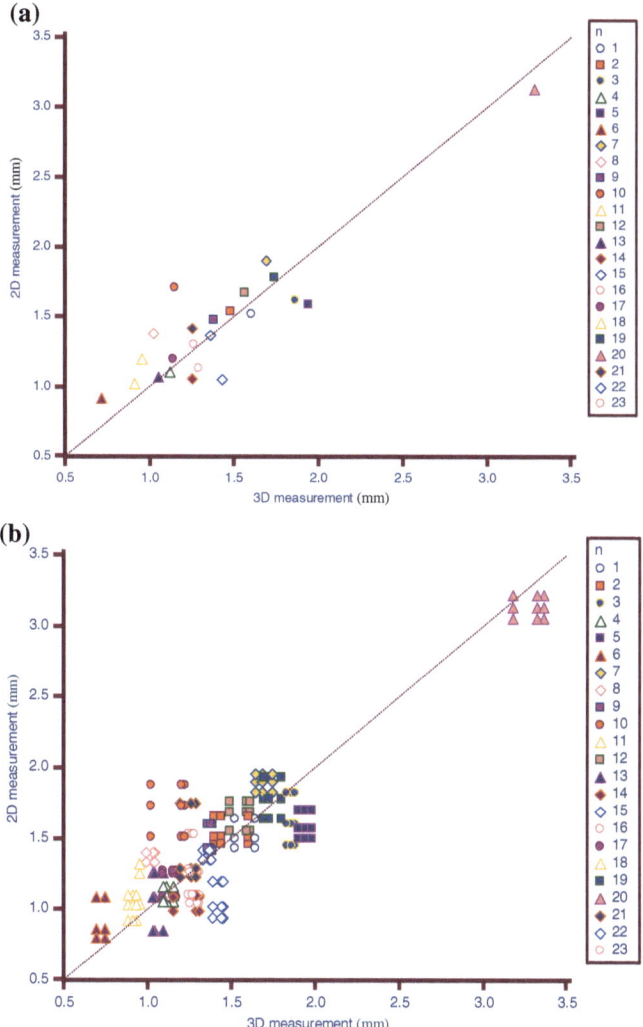

Fig. 4.4 Comparison of NT measurement two methods. **a** Mean 2D against mean 3D ($n = 23$). **b** 2D Trials against 3D trials ($n = 207$)

4.3 Intra Observer Variability

Total numbers of 23 clinician data are collected by two senior healthcare professional at hospital: denoted Operator A and Operator B. In order to answer the bias relative to intra observer variability, we need to investigate the repeatability skills of each operator before we can further to combine and analyze the collected data for numerical agreement. However, inter-observer variability analysis is excluded in this study since the prenatal screening and its measurements at the collaborated

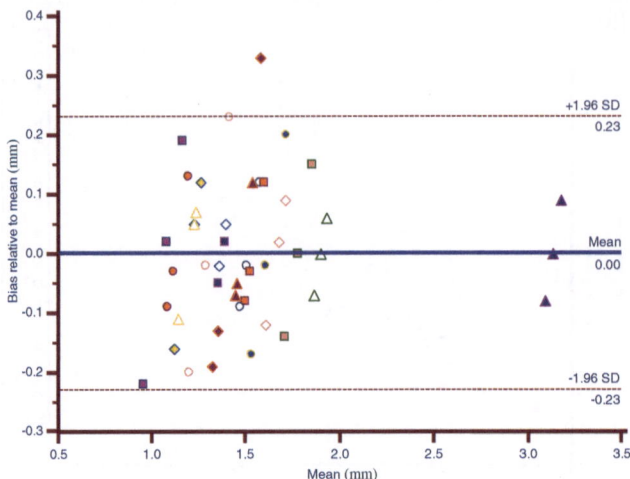

Fig. 4.5 Differences between trials A_i against mean $A(n = 48)$

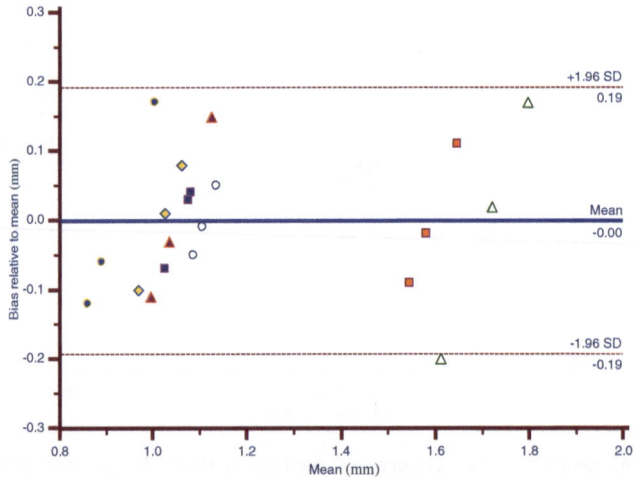

Fig. 4.6 Differences between trials B_i against mean B $(n = 21)$

hospital are not patient re-duplicated for each operator. Given each three successive trials per subject are measured by each operator: A_i and B_i; $i = 1, 2, 3$. We
can plot the difference of each individual trial to corresponding mean against mean
measurement, as shown in Fig. 4.5 and 4.6 below.

   Based on the observation above, one can noticed that the limits of agreements
for both operators are approximately overlapped; Operator A with sample size
equal to 48 shaped agreement $\pm 1.96\,SD$ equal to ± 0.22; Operator B with sample

size 21 shaped agreement $\pm 1.96\,SD$ equal to ± 0.19. Intra-observer agreement between operators did not show statistically significant differences for displacement measurements. Therefore, we can accept the assumption that both operators have same level of skills measurement for repeatability.

Although the case number was very small, and there are absences of inter-observer variability, our pilot study still demonstrated a novel 3D NT marker in first trimester screening for Trisomy 21. We will need more patient data collection which required long term collaboration with collaborated hospital and health-care professional in order to generate a standard curve measurement in the future. However, it will be very labor-expensive and time consuming to provide the data in every single case, especially in population based screening.

4.4 Level Agreement Between Measurements

We have repeated measurements three trials; $i = 3$, by two methods on the same individual subjects; $n = 23$. It is sensible to make use all of the three trial measurements for each method when conducting comparison analysis. There are three possible combination ways of observation for this investigation; first, each mean of the replicated measurement by each method on each individual is calculated, therefore, two pairs of means are used as input for Bland–Altman plot for limits of agreement analysis ($n = 23$); second, each corresponding individual trials of both methods are used as input ($3n = 69$); third, each possible combination trials for each individual subject corresponding to its method are used as input ($9n = 207$). Figure 4.7 shows the graphical representation of agreement Bland–Altman plot for first way observation analysis. The mean differences between 2D method measurements and 3D method measurements are plotted as y axis; against the average mean values of both methods as x axis.

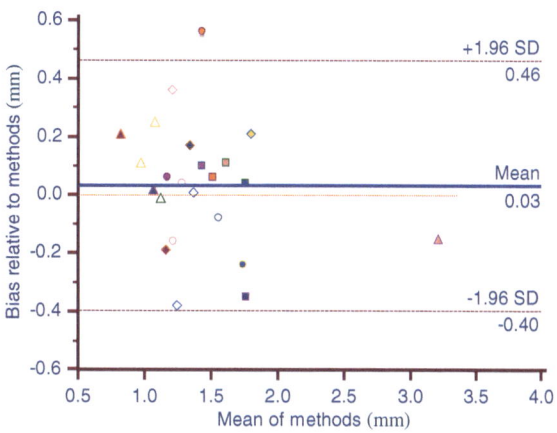

Fig. 4.7 Bias relative to methods against average mean methods ($n = 23$)

Based on the observation above, the arithmetic mean and standard deviation between methods is 0.03261 ± 0.2194 mm with 95 % CI from -0.06225 to 0.1275. By calculating the mean ± 1.96 SD, the upper limits of agreement is 0.4626 with 95 % CI from 0.2980–0.6271, the lower limit of agreement is -0.3973 with 95 % CI from -0.5619 to -0.2328.

The estimated bias of two methods will be unchanged by the mean computation (x-axis); however, the estimated standard deviation of the differences will be too small as the effect of measurement error has been removed. For that reason, correction of standard deviation s_d is needed. Denote the measurement of 2D and 3D methods as X and Y, the adjusted variance of difference between single measurements by each method is as follows (Bland–Altman 1999);

$$\widehat{\sigma}_d^2 = s_{\overline{d}}^2 + \left(1 - \frac{1}{m_x}\right) s_{xw}^2 + \left(1 - \frac{1}{m_y}\right) s_{yw}^2 \tag{4.1}$$

where, $\widehat{\sigma}_d^2 = Var(X - Y)$; $s_{\overline{d}}^2$ is the observed variance of the differences between the within-subject means; m_x and m_y is the number of replicated measurements, in our case equal to 3; s_{xw}^2 and s_{yw}^2 is the within-subject variance. From Table A, we can calculate the $\widehat{\sigma}_d^2 = 0.06209$, correction of standard deviation $s_d = \sqrt{\widehat{\sigma}_d^2} = 0.24918$,, so the adjusted 95 % limits of agreement is $0.0343 \pm 1.96\, s_d$ or from 0.5227 to -0.4541. The variance of adjusted limits of agreement can be estimated by follows;

$$Var\left(\overline{d} \pm 1.96\widehat{\sigma}_d\right) = \frac{\widehat{\sigma}_d^2}{n} + \frac{1.96^2}{2\widehat{\sigma}_d^2}\left(\frac{s_{\overline{d}}^4}{n-1} + \frac{(m_x - 1)\, s_{xw}^4}{nm_x^2} + \frac{(m_y - 1)\, s_{yw}^4}{nm_y^2}\right) \tag{4.2}$$

Where, $\overline{d}$ is the variance of mean difference. Hence, the standard error is $\sqrt{Var\left(\overline{d} \pm 1.96\widehat{\sigma}_d\right)} = 0.077429$. The 95 % CI for the lower limit of agreement is from -0.3023 to -0.6059; likewise, for 95 % CI of upper limit of agreement is from 0.6745 to 0.3709.

The second observation ways is to take considering each pairs of trials as independent measurement, so total $n = 69$, as plotted in Fig. 4.8 below. The bias measurement between each corresponding trials of the methods are plotted as y axis; against the average values of both methods trials as x axis.

The above arithmetic mean and standard deviation between methods is 0.03435 ± 0.2516 mm. By calculating the mean ± 1.96 SD, the 95 % CI upper limits of agreement is 0.5274 and the lower limit of agreement is -0.4587. No further correction of standard deviation in this observation way is required. Next, the third observation ways is by considering each possible combination of corresponding trials between the methods as the input for Bland–Altman plot analysis, total $n = 207$ Fig. 4.9.

The above arithmetic mean and standard deviation between methods is 0.03435 ± 0.2492 mm. By calculating the mean ± 1.96 SD, the 95 % CI upper limits of agreement is 0.5228 and the lower limit of agreement is -0.4541.

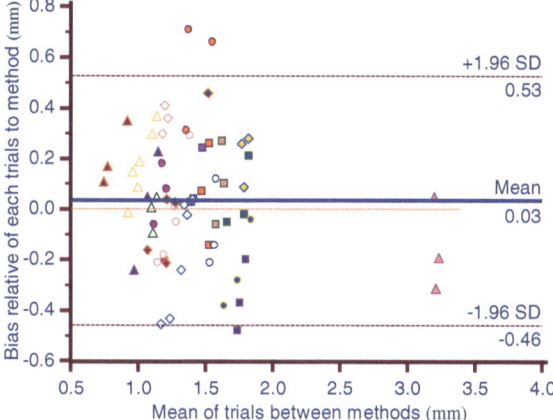

Fig. 4.8   Bias relative of each trial to methods against average methods trials ($n = 69$)

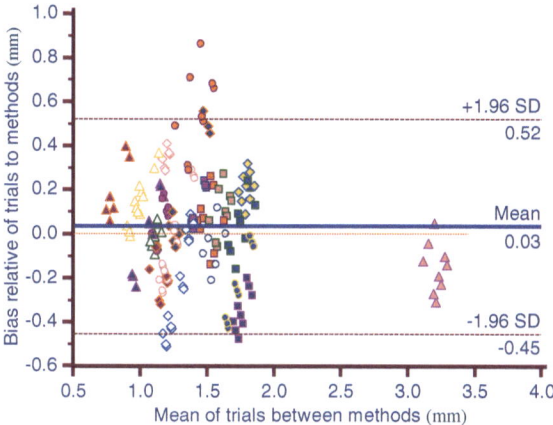

Fig. 4.9   Bias relative of each trial to methods against average methods trials ($n = 207$)

By comparing all the three observation ways of analyses above, all the arithmetic mean approximately remain the same at 0.03 values, meanwhile, almost all the markers are positioned within the 95 % limits of agreement which approximately overlapped in the same ranges of ±1.96 SD (Observation 1: 0.5227 to −0.4541; Observation 2: 0.5274 to −0.4587; Observation 3: 0.5228 to −0.4541). Limits of agreement between methods did not show statistically significant differences for displacement measurements. Also, one can noted that the results of adjusted limits of agreement and its standard deviation correction using Eq. 4.1 at Observation 1 matched exactly to the findings of Observation 3. This remarks that the 3D method measurement is highly agreed by the existing 2D method measurements.

Table 4.2 Paired samples t-test

Parameters	Statistics
Mean difference	−0. 03435
Standard deviation	0. 2492
95 % CI	−0.09422 to 0.02553
Test statistic t	−1.145
df	68
Asymp. sig	P = 0.2563

4.4.1 Paired Sample t-Test

On the other hands, we also analyzed the numerical mean differences between methods by using Paired sample t-test with hypothesis as follows;

H_0 : The difference between the means is equal to 0

H_a : The difference between the means is different from 0

where, H_0 is the null hypothesis, H_a is the alternative hypothesis. Table 4.2 below summarized the t-test statistics between the paired observations of the two group NT measurement together with the mean of the differences between the paired observations, and the standard deviation of these differences, followed by a 95 % confidence interval for the mean.

Given the computed P-value = 0.2563 (df. = 68) is higher than the significance level alpha = 0.05, one should failed to reject the null hypothesis H_0, and have no enough statistical evidence to accept the alternative hypothesis H_a. This concluded that the mean difference between the paired observations is statistically significant different equal to zero.

4.5 Methods Repeatability

Similar to the analysis of limits of agreement approaches using Bland and Altman's method, repeatability of each different method can be quantified for repeated measurement on the same subject obtained by the same method. The estimation of within-subject variance S_{xw}^2 (method 2D) and S_{yw}^2 (method 3D) can be estimated from the residual mean square (Bland and Altman 1999), using one-way analysis of variance; with subject (patient) as the factor. Then, we can compare the standard deviation of different method to investigate which method performs higher repeatability. The comparison of repeatability between both methods is relevant to method comparison because it limits the amount of agreement. In case of lack measurement repeatability of any methods, it may interfere the method comparison described above using limits of agreement. Therefore, this statistics is also useful to indicate a baseline against which to judge between-method variability. Figures 4.10 and 4.11 below shows the repeatability analysis for each method respectively using Bland–Altman plot, with total pairs of observation equal to 69.

Fig. 4.10 Total 2D trial-mean difference against 2D mean ($n = 69$)

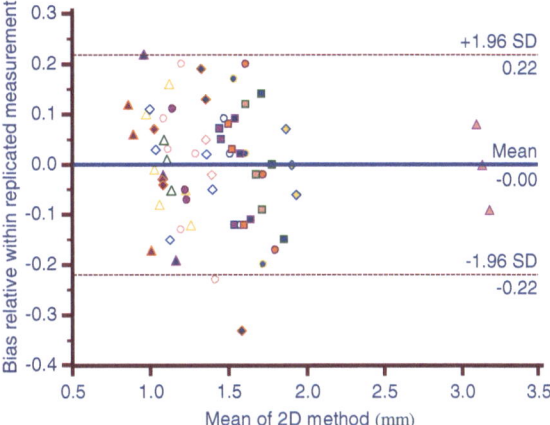

Fig. 4.11 Total 3D trial-mean difference against 3D mean ($n = 69$)

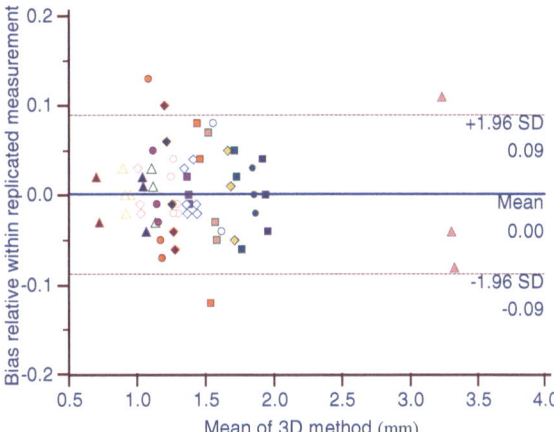

Bland–Altman plots above illustrate the three trial-mean differences of each method measurements against its mean for each replicated patient measurement. The ideal method's repeatability is expected the mean difference between the replicates measurement equal at zero; each replicate measurement on same individual have the same values, so that all the markers in the plots above shall lies on the mean line at value 0.00. Based on our findings, the standard deviation of replicated 2D measurement is 0.1116 with 95 % limits 0.2182 to −0.2193, where 3D measurements have 0.04506 with narrower 95 % limits 0.08949 to −0.08717 as compared to 2D measurements, which indicates that 3D measurements are closely disperse on the line equal to zero. Besides, returning to Table 4.1, we can compute the within-subject variance of 2D; S^2_{xw} equal to 0.018672. Likewise, for 3D method; S^2_{yw} equal to 0.003036. The *RC* can be calculated as follows;

$$RC = 1.96\, S_w \tag{4.3}$$

Therefore, the RC for 2D and 3D measurements are $RC_x = 0.26781$ mm and $RC_y = 0.10807$ mm. We can notice that the RC of 2D method is approximately 148 % greater than 3D method, which indicates 3D methods appears as more efficient method for higher repeatability.

4.6 Summary

Data distribution for collected clinician 2D method NT measurements are tested to be normally distributed with Kolmogorov–Smirnov test. Intra-observer variability is tested to be relatively low within Operator A and Operator B, indicates same skills level for repeatability within-subject measurement. The relationship among 2D and 3D measurements is modeled using Pearson product-moment correlation coefficient. Results revealed that 2D measurement is highly correlated with 3D measurement at R values equal to 0.8750 ($P < 0.0001$). Levels of agreement between methods are analyzed using Bland–Altman's test on three varied grouping observations. Limits of agreements among these grouped observations are found approximately intersecting each other with most of markers falls in ranges $\pm 1.96 \, SD$. Findings show that 3D measurement has higher repeatability with narrower 95 % CI limits of agreements, in addition, RC_X (2D) is found to be 148 % higher than RC_y (3D). However, paired samples t-test shows that there are no statistically significant mean differences between both tested methods.

4.7 Discussion

The fundamental aim of the developed 3D reconstruction and visualization program is to display the overall 3D volumetric fetal images, which provides comprehensive clinical image information for both internal and external body features; Considering the NT's tissue characteristics of fetus as an interior structure, composite function plays the key rule to separate NT from its background by integrating the rendering algorithms. The findings in previous section show that 3D measurement method has higher repeatability performance as compared to conventional 2D measurement method. One should note that there are five major improvements through 3D method yet there are no statistical significant mean differences between the measurements;

(1) The 3D measurements are considered as less human intervention and operator dependent. Unlike 2D B-mode ultrasound measurement, operator has to place the sonogram caliper manually on screen based on weak two echogenic lines for NT maximum thickness estimation. This requires highly trained and FMF competent operator with tedious manual measurement protocol. In 3D method, the NT structure is segmented using 3D seeded region inflation followed by interactive thickness measurement. Since the resultant 3D NT segmented structure are unequivocal from the ultrasound background images (dark pixels

in B-mode US images), NT marker will not be seen as appearing at the same visual rendering level with its surrounding voxels (dark voxels turned transparent), over-estimation or under-estimation are avoided inevitable.

(2) Undeniably, reconstructed 3D ultrasound fetal images have higher clinician values and diagnostics informative as compared to conventional 2D plane scans. The proposed method presented 3D reconstruction system from the conventional 2D US system at no costs while increasing the NT screening efficiency simultaneously. Volume data allows for a specific point in space to be evaluated from many different orientations by rotating, slicing, and referencing the slice to other orthogonal slices. It also allows for new volume-rendering displays that show depth, curvature, and surface images not available with conventional methods. This can increase the technology to patients' population benefits especially in limited budget hospital at developing countries.

(3) For instance, the best sagittal plane during the prenatal screening is favorable for physician to access the NT marker; this is of prime significance that current manual 2D scanning method is restricted to acquire the correct scanning plane of 2D fetal images; if the NT marker is not examined in an appropriate plane, the measurement would be shorter or longer than normal or in worst case—it does not exist. Also, if the tested images are not in the true sagittal view or coincide in the suitable plane, NT marker might not appear in appropriate position. Based on current 3D findings, physician is aware of their selected NT position. Precise NT thickness measurements are feasible within 3D Euclidean distance measurement.

(4) Repeatability within subjects' measurement from the proposed method is promising as compared to conventional 2D method. This repeatability is highly relevant to the limits of agreements analysis shown in previous section. High level of agreements but lack of repeatability would leads to poor precision since there are considerable variations in repeated measurements on the same subjects.

(5) The proposed NT's profile visualizing and measurements method are not restricted by the fetus position. Contrariwise, B-mode 2D NT ultrasound visualizing is dependent on arbitrary scanning view. In cases where the fetal position did not allow optimal visualization of the fetal neck in 2D view, NT was easily obtained in 3D model regardless of fetal position. The obtained 3D images could then be manipulated to get all the desired views for NT measurement. The potential advantages in clinical practice are pivotal: diagnostic time reduction and exceptional volume data storage; allowing the measurements and anatomical survey to be repeated at any time.

4.7.1 Advantages and Disadvantages of Early Fetal Abnormalities Detection

Although the FMF recommended 12–14 weeks prenatal scan protocol is effective in detecting a large variety of fetal abnormalities, yet, there are also some potential advantages and disadvantages cross over with second-trimester ultrasound examination. The earlier detection mentioned in this book refer to first trimester of pregnancies, mainly before 14 weeks of ultrasound prenatal screening. The advantages of earlier detection of fetal abnormalities allowed patients and doctors to make earlier decision on further pregnant-healthcare management. In case of severe fetal defection that might endanger both baby and mother's life, termination of pregnancy has to be decided as early as possible. Early termination is supposed to be both medically and psychologically advantageous. Patient's privacies are protected and potential harms to mother's body are eliminated to the lowest prospects. In general, early pregnancy termination for fetal abnormalities is more acceptable than late termination. Besides, first trimester ultrasound scanning provides reassurance of normality in early pregnancy is particular useful to women who having abnormal detection from previous pregnancies. A detailed morphology scan and NT examination at first trimester are expected to relieve the anxiety of patients especially those have babies affected by genetic anomalies including Trisomy 21 from previous pregnancies. However, the medical experts suggested that even the fetus with a confirmed increased nuchal translucency have normal karyotype tests; he or she should also have detailed scanning at 12–14 weeks for major structural abnormalities examination, and followed by the 16–20 weeks follow-up examination by conventional second trimester.

Although the majority of the fetal abnormalities can be diagnosed in early pregnancy, the disadvantages of early detection may potentially result in termination of an otherwise normal pregnancy. By excluding the human factor where patient who want to have perfect healthy babies, the small size of fetal examination marker are limiting the factor in obtaining optimal sonographic visualization. The sensitivity of ultrasound scanning to detect fetal anomalies is also related to the operator's experience and the resolution of the ultrasound machine. This exploration is more time-consuming and requires a high level of training of the examiner. However, without first trimester scan, some of the chromosomal abnormalities associated with cystic hygroma may have been missed because cystic hygroma, even when associated with chromosomal abnormalities, usually resolve after 14 weeks.

References

Bland, J. M., & Altman, D. G. (1986). Statistical-methods for assessing agreement between two methods of clinical measurement. *Lancet,* *1*(8476), 307–310.

Bland, J. M., & Altman, D. G. (1995). Comparing methods of measurement—why plotting difference against standard method is misleading. *Lancet,* *346*(8982), 1085–1087.

Bland, J. M., & Altman, D. G. (1999). Measuring agreement in method comparison studies. *Statistical Methods in Medical Research,* *8*(2), 135–160.

Bland, J. M., & Altman, D. G. (2007). Agreement between methods of measurement with multiple observations per individual. *Journal of Biopharmaceutical Statistics,* *17*(4), 571–582.

Chapter 5
Future Improvements

Abstract This chapter provides the conclusion for the system testing and evaluation. It also gives some recommendation for further improvement of the system.

5.1 Conclusions

This book has reviewed related methods and algorithms for fetal abnormalities detection and proposed a very first version of open-sources three dimensional nuchal translucency for fetal abnormalities detection based on ultrasound synbook. The 3D reconstruction volume data allow evaluation of a specific point in space from many different orientations by rotating, slicing, and referencing the slice to other orthogonal or arbitrary slices. It also allows for new volume rendering displays that show depth, curvature, and surface images which are not available with conventional 2D methods. Hence, implementation of composite function using color and opacity characteristic setting in 3D models has demonstrated improvement in nuchal translucency (NT) identification. The interest image features are visually operator-intuitive in 3D rendering scene, provides outstanding visualization and better ultrasound marker evaluation as compared to conventional B-mode method. Semi-automated 3D seeded region based segmentation technique allow the ROI inflation within the two fold layer of nuchal translucency. It is suitable for single region inflation as nuchal translucency formation is the accumulation of fluid beneath fetal neck that forms an isolated homogenous region in ultrasound images. We have shown that this technique can improved the measurement substantially with higher repeatability performance while reducing human intervention which includes 2D mid-sagittal plane selection and tedious protocol for manual 2D caliper placements. Paired t-test shows there are no statistical significant mean difference between the paired observations but 3D method has higher repeatability; $RC_y = 0.10807$ (3D) compared to $RC_x = 0.26781$ (2D). As expected, both methods have high degree of association, Pearson's correlation coefficient, $R = 0.8750$, $P < 0.0001$. Bland–Altman test reveals that 3D method has significant correspondence with 2D method with $\mu \pm 95\,\%$ CI limits of agreement: 0.5227 to -0.4541. This can further benefit the non-expert operators without taking the FMF competency training. The bias relatives of differences between and within operators are therefore improved. It is expected the small improvement of NT measurement can improved the overall screening performance (Wright et al. 2008). The observed findings have statistically demonstrated the feasibility of proposed 3D ultrasound nuchal translucency evaluation method.

The novel contribution of the research is that a new method for Trisomy 21 early detection using 3D ultrasound approach has been developed. This new method has an edge over the conventional 2D method in terms of its repeatability of performance, less reliance on human intervention, and reduction of inter and intra observer variability.

5.2 Limitations and Recommendations

There are several challenges that have yet to be tackled for general acceptance of the technology. To acquire the most accurate NT assessment, there are two vital keys to be complied; acquisition and calipers placement for measurements. Intra- and inter-operator error and measurement variability are not only exposed to the fault calipers placement, but it is also subjected to the concern of imprecise image acquisition. In this book, we presented a complicated process for accurate NT measurement, which is mainly the later key described above. The benefit of the proposed semi-auto 3D NT segmentation and marker evaluation lies in the partial standardization of the NT image acquisition processes developed by FMF. Acquisitions itself, however, still remained as the limitation that are much more operator-dependent, and yet cannot be performed automatically. The resulting 3D ultrasound reconstruction images are based on the acquired 2D ultrasound images. Consequently, the 3D results are heavily dependent on the accuracy of acquired 2D image. Unfortunately, there are no feasible ways yet to replace human handson ultrasound scanning method of medical doctor using artificial automated technique. The recommendation for the potential future works so as to support future clinical studies consists of the following majors.

5.2.1 The Need of Scanning Systems Improvement

The existing commercial 3D scanning systems are operated at high frequencies ultrasound probe, ranges from 8 to 12 MHz, which are not appropriate for distinct depth resolution of NT assessment. Slow data acquisition can also results the blurred 3D images due to the fetal movement. Technological limitations for data storage and manipulation in real-time 3D ultrasound need to be overcome so to provide an intuitive interface for both expert and non-expert operators. The recent technique proposed to encounter the 3D ultrasound motion artifacts is the gated electrocardiogram (ECG) during ultrasound scanning. It aims to acquire the collectivity of B-scans at the same point in the cardiac cycle before 3D ultrasound reconstruction. This technique has shown feasibility for cardiology ultrasound scanning and eliminated the heart motion; however, fetal movement is not an easy case here. So far there has no feasible way to control the fetal movement at the moment while acquiring B-scans; this is the reason that hands-on screening technique of medical doctors is not replaced yet, as there are too many parameters to be controlled.

5.2.2 The Need of Image and Position for Freehand Acquisition Improvement

As described in section acquisition in Chap. 3. Several position tracking protocols are introduced for image-coordinate synchronization before 3D ultrasound image reconstruction. Each of the protocol has their advantages and its limitations that precluding their use in common clinical environment such as metallic magnetic field interferences, optical light obstacle, bulky mechanic facilities and others. For this reason, this explained that all the freehand scanning protocols are yet to be predominated used for research studies under carefully controlled environment. In future, perhaps hybrid protocol might be a potential method for better acquisition method without losing its nature advantages. It is low costs and appears as higher flexibility for 3D volume scanning if compared to existing commercial 3D probes. It is not limited by the mechanic housing of the probe that framed with its coordinate references, so it is able to scan object of interest which are bigger than the size of probe housing.

5.2.3 The Need of Algorithms for Fully Automated Segmentation

Fully automated 3D segmentation will be the advanced computing technique and tend to be the trend in future clinical computing studies. Dynamic parameters including metric function and seed selection are expected to become hidden computing layer, thus reducing factor that may prone to operator-dependent variation. However, it should be carefully handled to avoid misuse or abuse of fully automation. The operators should to be fitted with expected clinical background knowledge and may not overreliance on the automated findings. They should able to review the results, and whenever necessary, discard the fault findings or unexpected measurement and rerun the algorithm to avoid automation blasphemy.

References

Wright, D., Kagan, K. O., Molina, F. S., Gazzoni, A., & Nicolaides, K. H. (2008). A mixture model of nuchal translucency thickness in screening for chromosomal defects. *Ultrasound in Obstetrics and Gynecology, 31*(4): 376–383.

Glossary

Table A Nuchal translucency measurement made simultaneously on each individual by two methods with each making three observations

n	2D			3D		
	A_1	A_2	A_3	B_1	B_2	B_3
P1	1.64	1.43	1.50	1.52	1.64	1.64
P2	1.46	1.66	1.51	1.6	1.4	1.44
P3	1.45	1.60	1.82	1.83	1.88	1.86
P4	1.16	1.10	1.06	1.11	1.09	1.15
P5	1.50	1.70	1.57	1.98	1.9	1.94
P6	1.09	0.86	0.80	0.74	0.69	0.69
P7	1.90	1.83	1.96	1.64	1.74	1.68
P8	1.33	1.40	1.40	1.03	0.99	1.04
P9	1.60	1.43	1.41	1.36	1.39	1.38
P10	1.73	1.88	1.51	1.02	1.22	1.2
P11	1.04	1.25	1.32	0.95	0.95	0.95
P12	1.55	1.76	1.69	1.61	1.49	1.59
P13	0.85	1.09	1.26	1.09	1.04	1.03
P14	0.99	1.09	1.10	1.15	1.29	1.31
P15	1.20	0.94	1.02	1.44	1.39	1.45
P16	1.10	1.04	1.26	1.3	1.25	1.31
P17	1.09	1.27	1.25	1.15	1.09	1.17
P18	0.92	1.03	1.10	0.93	0.88	0.91
P19	1.78	1.64	1.93	1.8	1.69	1.72
P20	3.05	3.13	3.22	3.36	3.32	3.17
P21	1.75	1.23	1.29	1.29	1.19	1.26
P22	1.35	1.35	1.42	1.33	1.37	1.38
P23	1.53	1.28	1.10	1.24	1.27	1.28

K. W. Lai and E. Supriyanto, *Detection of Fetal Abnormalities Based on Three Dimensional Nuchal Translucency*, SpringerBriefs in Applied Sciences and Technology, DOI: 10.1007/978-981-4021-96-8, © The Author(s) 2013